GUIDE TO SOURCES IN NORTHERN CALIFORNIA FOR HISTORY OF SCIENCE AND TECHNOLOGY

Berkeley Papers in History of Science

X

GUIDE TO SOURCES IN NORTHERN CALIFORNIA FOR HISTORY OF SCIENCE AND TECHNOLOGY

Robin E. Rider and Henry E. Lowood

Office for History of Science and Technology
University of California, Berkeley

1985

This volume was prepared for distribution
at the XVIIth International Congress of
History of Science, University of California,
Berkeley, 31 July – 8 August 1985.
Copies are available through the
Office for History of Science and
Technology.

Copyright © 1985 by The Regents of
the University of California.

ISBN 0-918102-12-X

ISSN 0145-0379

Library of Congress catalog
card no. 85-51383

*

CONTENTS

Introduction — i

Berkeley — 1

    Be.01 Lawrence Berkeley Laboratory Archives — 5

    Be.02 University of California Archives — 6

    Be.03 Office for History of Science and Technology — 12

    Be.04 Judah L. Magnes Museum — 14

    Be.05 Bancroft Library — 15

    Be.06 Water Resources Center Archives — 84

Davis, University of California (Da.01) — 85

Sacramento — 97

    Sa.01 California State Archives — 97

    Sa.02 California State Library — 98

San Francisco — 101

    SF.01 California Academy of Sciences — 101

    SF.02 University of California — 102

Santa Cruz	107
SC.01 Lick Observatory Archives	107
SC.03 University of California	111
Stanford	113
St.01 University Archives	115
St.02 Special Collections	132
St.03 Hoover Institution Archives	138
St.04 Stanford Linear Accelerator Center	147
St.05 SRI International	148
St.06 Lane Library	149
List of oral histories	164
Index	166

INTRODUCTION

Beginning with the Spanish expeditions of the 16th century, Europeans explored the natural wealth of California—its inviting harbors, its fruitful union of mild climate and fertile soil, the precious gold beneath the surface, the towering trees above it. The discovery and exploitation of these riches are documented in the libraries and archives of northern California. Their collections of published and unpublished materials reflect the historical and cultural diversity of California from the establishment of Spanish missions through the advent of Forty-Niners eager for gold to the emergence of the state as a potent economic and political force. Surveyors and solitary wanderers investigated the geology of the Sierra Nevada; archæologists and ethnographers struggled to preserve the culture of California Indians menaced by the imperatives of expansion and change. Journals and notes preserved in archives provide a record of these investigations and constitute valuable sources of field data. The quickening pace of settlement, encouraged by the railroad, also occasioned the founding of universities, including the University of California in Berkeley (1868) and Stanford University on the San Francisco peninsula (1891). Their histories, documented in institutional archives and the papers of their faculties, illustrate the combination of private philanthropy and public support, industry and academe, scientific innovation and entrepreneurial skill, that created important centers of research and development in the San Francisco Bay Area.

Introduction

This *Guide* is intended to lead researchers interested in the development of science and technology to relevant manuscript collections at a few prominent institutions in northern California.[1] Many of the holdings of these and other California repositories are reported in union catalogs, including the *National union catalog of manuscript collections*, and national databases. Some of the collections described here are also represented in the international *Inventory of sources for history of twentieth-century physics*, which may be consulted at the Office for History of Science and Technology, University of California, Berkeley.

Entries in this *Guide* are organized by location, and each reporting library or archive is assigned a code according to the conventions of the physics *Inventory*. For most of the repositories, manuscript holdings pertaining to science and technology are listed alphabetically by the collection titles used at the library or archive in question. Brief descriptions of individual collections provide general information about size, scope, and content, and availability of finding aids. In some instances, shelf-list numbers and restrictions on access are indicated. Researchers interested in individual collections should write or call the repository; addresses and telephone numbers are listed at the beginning of each section of the *Guide*. Retrieval of collections from remote storage may require a week's notice; special arrangements are generally needed for access to collections that have not been fully processed.

An index at the end of the *Guide* provides access by subject and name. Index entries lead the reader to the repository (via the archive code) and, when appropriate, to an individual collection name. An alphabetical list of all oral histories included in the *Guide* is also appended.

---

1. Other California repositories of interest are listed in *Archival and manuscript repositories in California*, compiled by the staff of the California State Archives (Society of California Archivists, 1984); they include the archives and library of the California Academy of Sciences in San Francisco, the Electronics Museum–De Forest Memorial Archives in Los Altos, and the U.S. Federal Archives and Records Center in San Bruno.

Introduction

Codes and abbreviations

UC	University of California
UCB	University of California, Berkeley
UCD	University of California, Davis
UCSC	University of California, Santa Cruz
UCSF	University of California, San Francisco
Be.01	Archives, Lawrence Berkeley Laboratory (formerly the Radiation Laboratory), UCB
Be.02	University Archives, UCB
Be.03	Office for History of Science and Technology, UCB
Be.04	Judah Magnes Museum, Berkeley
Be.05	Manuscripts Division, The Bancroft Library, UCB
Be.06	Water Resources Center Archives, UCB
Sa.01	California State Archives, Sacramento
Sa.02	California State Library, Sacramento
SC.01	Lick Observatory Archives, UCSC
SC.03	University Library, UCSC
SF.01	California Academy of Sciences, San Francisco
SF.02	Special Collections, University Library, UCSF
St.01	University Archives, Stanford University
St.02	Manuscripts Division, University Library, Stanford
St.03	Hoover Institution, Stanford
St.04	Stanford Linear Accelerator Center
St.05	SRI International (formerly Stanford Research Institute)
St.06	Lane Medical Library, Stanford

Acknowledgments

The *Guide* has been produced in conjunction with the XVIIth International Congress of History of Science, held on the Berkeley campus of the University of California from 31 July to 8 August 1985. Funding was provided by the International Congress for History of Science, the Office for History of Science

Introduction

and Technology, and The Bancroft Library (UCB). It is a pleasure to acknowledge the cooperation of Sabra Basler (UCD), Geraldine Davis (California State Library), Vicki Davis (Lawrence Berkeley Laboratory Archives, UCB), Margaret Felts and Dorothy Schaumberg (UCSC), Lynn Fonfa and Jane Levy (Judah L. Magnes Museum), Gerald Giefer (Water Resources Center Archives, UCB), Johan Kooy (California Academy of Sciences), Dave Snyder (California State Archives), and Nancy Zinn (UCSF), who provided descriptions of holdings at their institutions. Special thanks are due to our colleagues at The Bancroft Library and the Stanford University Libraries for their encouragement, and to J. L. Heilbron, Bruce R. Wheaton, and the staff of the Office for History of Science and Technology for their advice and assistance.

Robin E. Rider
University of California, Berkeley

Henry E. Lowood
Stanford University

15 June 1985

GUIDE TO SOURCES

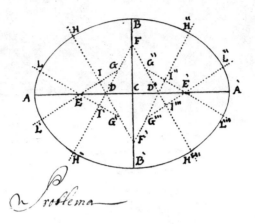

Problema

Dati gli apsi $AC\grave{A}$, $BC\grave{B}$ di un ovale, descriverla con degli archi circolari ciascuno di 30 gradi del suo circolo, che nelle unioni abbiano le tangenti communi.

Eighteenth-century mathematical manuscript.
Bošković papers, The Bancroft Library.

BERKELEY

Libraries, archives, and research units on the Berkeley campus, the first of nine campuses now constituting the University of California, offer varied resources to historians of science and technology. Among the six million volumes in the campus library system are extensive holdings in science and technology, including many complete runs of scientific periodicals dating back to the earliest years of European and American scientific academies. Thanks in part to the collecting efforts of such Berkeley professors as Florian Cajori and Charles Kofoid (see Be.05 below), UCB libraries are fortunate to possess a strong collection of classic works in early modern and modern science; these titles are to be found in Doe Library, The Bancroft Library, and branch libraries in scientific fields. The Bancroft Library, repository for most special collections in the Berkeley library system, has continued, as part of the activities of its History of Science and Technology Program, to acquire books both by lesser-known authors and by scientific luminaries. Current efforts to build the Bancroft collection of rare books emphasize the physical sciences, mathematics, and scientific instruments in the 18th century; newly acquired textbooks and specialized research publications in these areas complement extensive holdings in other fields and periods. The Biology Library is especially rich in rare and valuable books, often lavishly illustrated, in the life sciences and medicine. Union catalogs, in both card and machine-readable form, provide access to books and serials in Doe Library, the Bancroft, and branch libraries on the Berkeley campus. Catalog information can be obtained by telephone: (415) 642-6657.

Archival and manuscript resources available at Berkeley also relate to diverse fields of science and technology. Many are located in The Bancroft Library, which houses the University Archives as well as an extensive manuscript division. The Bancroft builds on its strong holdings by acquiring manuscript collections reflecting all phases of the development of the American West. These collections, which range from the papers of early explorers to those of contemporary California authors, include the correspondence and papers of eminent scientists and engineers, acquired under the auspices of the History of Science and Technology Program (HSTP).

The mandate of the HSTP is to document the growth of science and technology by means of manuscript collections, oral history, and published works. Particular strengths in the holdings assembled by the Program are the 20th-century physical sciences, nuclear medicine, electronics on the San Francisco Peninsula, and anthropology. Also notable are the correspondence and papers of the 18th-century Jesuit natural philosopher Rudjer Josip Bošković. A rich collection of his many publications, together with manuscripts of his writings in science, philosophy, and belles lettres and his correspondence with savants throughout Europe, offers valuable insights into scientific thought and the workings of scientific institutions in the 18th century.

To complement manuscript holdings, Program staff have conducted interviews focused on such topics as nuclear science at the Radiation Laboratory (later Lawrence Berkeley Laboratory), including the research in medical physics at Donner and Crocker Laboratories; the connections of science, technology, and business in the development of radio and electronics in northern California; the influence of physics research in industry; and the growth and application of operations research in California.

The extent of Bancroft manuscript collections and oral histories is suggested by the hundreds of individual entries below, selected for their possible interest to historians of science and technology. Many other collections at the Bancroft, not listed separately, deal with exploration, mining, agriculture, communication and transportation systems, conservation, and land use, as well as politics and government, literature and the arts, economics and business. The Bancroft Library's pictorial, map, book arts, and papyri collections will also be of interest for researchers investigating the history of science and technology.

The remainder of this section describes archival resources for science and technology available at the Office for History of Science and Technology, Lawrence Berkeley Laboratory Archives, and Water Resources Center Archives on the Berkeley campus, and the Judah L. Magnes Museum of Berkeley.

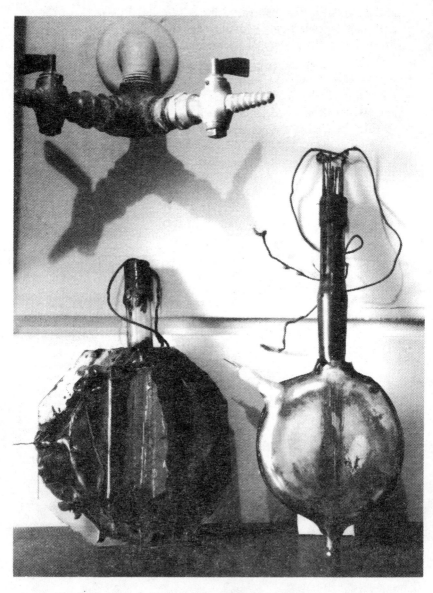

Accelerating "tubes" of the first cyclotron (1930). University Archives, The Bancroft Library.

Be.01

Archives and Records Office
Lawrence Berkeley Laboratory
Building 69, Room 201
University of California
Berkeley, CA 94720
(415) 486-5525

The Archives and Records Office of Lawrence Berkeley Laboratory contains approximately 13,000 cubic feet of records relating to historical and current operations of the Laboratory. The Radiation Laboratory was founded in 1931 by Ernest O. Lawrence on the Berkeley campus of the University of California. The Laboratory, now located in the hills above the campus, is funded by the federal government and operated by the University of California.

Archival collections include business records of the Manhattan Engineer District (1942–1945); records of the Office of the Director under Edwin McMillan and Andrew Sessler (1958–1980); records of the Associate Division for Administration (1938–1978), including the papers of Donald Cooksey; and scientific records from the Physics, Biology and Medicine, Accelerator, Energy and Environment, Nuclear Science, and Earth Sciences divisions. Because the records are housed in several locations, it is advisable to make advance arrangements to consult the collections.

Be.02

University Archives
The Bancroft Library
University of California
Berkeley, CA 94720
(415) 642-2933

The papers and memorabilia preserved in the University Archives at Berkeley pertain both to the Berkeley campus, the original site of the University of California, and to the entire UC system of nine campuses. These materials may be used in the Heller Reading Room of The Bancroft Library. Advance notice is encouraged.

Antivivisection initiatives.
> CU-381. 1 box (ca. 1920–1922). Correspondence, clippings, statements, printed materials, etc., concerning proposed legislation.

Childs, Herbert.
> CU-369. 4 boxes. Correspondence, notes, and research material relating to his biography of Ernest O. Lawrence.

Fraenkel-Conrat, Heinz Ludwig (1910–   ), professor of molecular biology, UCB.
> Archives Tape C:30. Recording of his faculty research lecture, "Viruses as research tools in molecular biology" (1968).

Fuller, George.
> CU-400. Slide rule (1879).

Hildebrand, Joel Henry (1881–1983), professor of chemistry, UCB.
> Archives Tapes C:67, 71. Recordings of his 100th birthday luncheon and convocation (1981).

Hitch, Charles Johnston (1910–), university president, UC.

CU-303. 4 boxes (1968–1975). Correspondence and papers as university president.

Kerr, Clark (1911–), chancellor, UCB; university president, UC.

CU-302. 37 boxes (1952–1957). Correspondence and papers as chancellor of UCB and as university president.

Lawrence, Ernest Orlando (1901–1958), professor of physics, UCB; Nobel laureate.

Archives Discs: 7, 15. Recordings, including one of the Nobel Prize ceremony in Berkeley (1940).

Lenzen, Victor Fritz (1890–1975), professor of physics, UCB.

CU-5.547. 1 box (1948–1955). Correspondence and papers concerning the Administrative Committee for Lick-Wilmerding School.

Reagan, Ronald (1911–).

Archives Tape C:21. Recording of remarks during a visit of the UC Regents to Los Alamos Scientific Laboratory (1967), while Reagan was governor of California.

Research Corporation.

CU-400, env. 39, 40. Scientific awards given to Ernest O. Lawrence (1937) and Edwin M. McMillan (1950).

Sproul, Robert Gordon (1891–1975), university president, UC.

CU-301. 57 boxes, 13 v. Personal correspondence and papers as university president and president emeritus, including scrapbooks of clippings.

Teller, Edward (1908–), professor of physics, UCB.

Archives Disc: 19. Recording of "The size and nature of the universe: The theory of relativity."

Thacher, Edwin.

CU-400. Calculating devices (ca. 1880–1884).

Turner, Francis J. (1904–), professor of geophysics, UCB.

Archives Tape C:32. Recording of his faculty research lecture (1970).

Underhill, Robert McKenzie (1893–), secretary of the UC Regents.

CU-386. 1 box (1952–1953), unsorted. Papers relating to his membership on the Department of Defense Research and Development Board.

U.S. Work Projects Administration.

CU-5.46. 3 boxes (1936–1940). Records of UC projects.

University of California. Academic Senate.

CU-9. Records (1869–), finding aid. Includes records of the Committee on Research (1916–1967), papers concerning the loyalty oath, etc. Among the chairmen of Senate committees were Wendell Stanley and Joel Hildebrand.

——. **Academic staff biographical cards.**

CU-10. 15 drawers (1869–1958).

——. **Agriculture, College of.**

CU-20. Correspondence and papers, including records of experimental stations and the correspondence and papers of Frank Adams. Some papers relate to irrigation and to viticulture.

——. **Anthropology, Department of.**

CU-23. 61 cartons (1901–1956), finding aid. Correspondence and records, including some records of the Museum of Anthropology.

——. **Anthropology, Museum of.**

CU-23.1. 32 boxes, finding aid. Ethnological documents.

——. **Appropriate Technology Program.**

CU-383. 1 box (1977–1984). Background correspondence and other materials, including minutes of the advisory committee and research project abstracts.

———. **Architects and Engineers.**

CU-13. Plans, reports, building files, specifications, etc. Includes papers concerning the construction of UC laboratories.

———. **Astronomy, Department of.**

CU-25. 39 boxes (1882–1960), finding aid. Department records.

———. **Budget Office.**

CU-3.2. Records (1912–1964), including departmental budget requests. Also CU-4.fin.m, containing minutes of meetings of the finance committee of the Regents.

———. **Chancellor, Office of the.**

CU-150. 4 boxes (1966–1981). Chancellor's directives. Also the papers of the chancellor's office under Clark Kerr, Glenn T. Seaborg, and Edward Strong (ca. 1952–1961) in CU-149.

———. **Chemistry, College of.**

CU-30. 10 boxes, 1 v. (1874–1955), finding aid. Correspondence, notes, subject files, etc., of faculty members, including Edmond O'Neill, G. N. Lewis, William Bray, Wendell Latimer, John I. Winkler, and Willard B. Rising; minutes of faculty meetings (1909–1916).

———. **Donner Laboratory.**

CU-68.4. 1 box (1942–1946). Records of the aeromedical laboratory.

———. **Engineering, College of.**

CU-39. 11 boxes (1906–1954). Records.

———. **Engineering, Science, and Management War Training Program.**

CU-354. 7 boxes (ca. 1940–1945). Records of Program activities at UCB.

———. **History of Science Dinner Club.**

CU-340. 3 boxes (1933–1970). Records of the organization founded by Herbert M. Evans.

——. **Lawrence Hall of Science.**

CU-49. 3 boxes (1958–1979), finding aid. Records.

——. **Mining Association.**

CU-211. 7 v. (1902–1960). Minute books. See also CU-202.5.

——. **Patent Office.**

CU-6. 1 carton (1962–   ). Records of the systemwide university patent office. Also CU-5, 1398–1424.

——. **Physics, Department of.**

CU-68. 8 boxes (ca. 1920–1962). Departmental records.

——. **President, Office of the.**

CU-5. 1464 cartons (1886–1979), card index. Correspondence and papers. Includes much material relating to science departments, laboratories, and faculty.

CU-5.1. 5 boxes (1930–1942). Reports to the university president from academic departments (at UCB and UCLA), branches, and administrative offices.

——. **Public Health, School of.**

CU-71. 1 box (1932–1943). Correspondence relating to the curricula in nursing.

——. **Radiation Laboratory.**

CU-68.3. 1 box (April 1943–June 1945). Records of telephone conversations.

——. **Regents.**

CU-1. 97 boxes (1868–1933), card index. Correspondence and papers of the Office of the Secretary to the Regents. Includes 3 volumes of letters to Lick Observatory (1888–1898) in Cu–1.Lick, and 23 boxes pertaining to contracts (1941–1969) in CU-1.8.

——. **Sponsored Projects Office.**

CU-151. 80 cartons (1971–1979), finding aid. Records.

Student notes and notebooks.
CU-300.

———. **Vertebrate Zoology, Museum of.**
CU-120. 5 boxes (1908–1949), finding aid. Correspondence and papers, including correspondence between Annie M. Alexander and museum directors.

Be.03

Office for History of Science and Technology
470 Stephens Hall
University of California
Berkeley, CA 94720
(415) 642-4581

The Office for History of Science and Technology (OHST) coordinates activity concerning history of science in the northern California region and constitutes the research and resource center for history of science and technology at Berkeley. It is closely associated with the instructional program of the Department of History through shared faculty and student research assistants. The OHST administers research projects in history of science and technology and extends to visiting scholars affiliated with it access to the extensive library holdings of the University.

The Office provides a place where students and faculty with common research interests meet frequently and informally. It sponsors colloquia and symposia in many areas of history of science, compiles and publishes bibliographies and other research tools as part of the series *Berkeley papers in history of science*, and, in cooperation with the University of California Press, edits and publishes *Historical studies in the physical sciences*. Visitors from other institutions working in history of science are welcome to use the facilities of the Office while in Berkeley. Inquiries concerning affiliation may be addressed to the director, Professor J. L. Heilbron.

In conjunction with The Bancroft Library, the Office for History of Science and Technology is one of several depositories of the Archive for History of Quantum Physics, a collection of microfilmed documents and interviews concerning the development of the quantum theory of the atom in the period 1896–1930. The approximately 100 reels of microfilm reproduce over 25,000 items of correspondence and unpublished manuscripts of several hundred physicists; their contents are described in T. S. Kuhn, J. L. Heilbron, Paul Forman, and Lini Allen, *Sources for history of quantum physics: An inventory and report* (Philadelphia, 1967).

The Office has also undertaken an extensive search for documentation of the history of 20th-century physics. The systematic results of that search constitute the *Inventory of sources for history of twentieth-century physics*, now available to researchers at the OHST. The *Inventory* contains detailed archival information assembled from over a thousand repositories in thirty-five countries, which can lead researchers interested in the evolution of recent physics to more than 500,000 letters containing contemporary historical documentation of the field from 1896 to the middle 1950s. The *Inventory* has been entered into a computer database especially designed for this purpose, permitting sophisticated searches and compilation of data.

Subsidiary results of the *Inventory* have been published in the series *Berkeley papers in history of science*, available from the OHST. These include *Literature on the history of physics in the twentieth century*, containing over 7,000 references, and *An inventory of published letters to and from physicists, 1900–1950*, listing on microfiche more than 25,000 quoted letters. The series also includes bibliographies of the non-technical writings of W. H. and W. L. Bragg, Max Planck, and Ernest Rutherford; an extensive bibliography of the writings of Werner Heisenberg; a bibliography of quantitative studies on science and history of science; a detailed calendar of the correspondence of Pierre Simon Laplace; and a bibliography of algebra from the beginning of printing until 1800.

Also available at the OHST are noncirculating biographical and bibliographic reference works in history of science, runs of relevant historical journals, and microfilm copies of correspondence and papers in other archives and libraries. Among the latter are microfilms of correspondence and manuscripts of Niels Bohr, Paul Ehrenfest, H. A. Lorentz, H. G. J. Moseley, Ernest Rutherford, Owen Richardson, and Karl Schwarzschild.

Be.04

Judah L. Magnes Museum
The Jewish Museum of the West
2911 Russell Street
Berkeley, CA 94705
(415) 849-2710

Both the Western Jewish History Center and the Blumenthal Library at the Magnes Museum contain manuscript collections relating to science and technology. Among the holdings of the Western Jewish History Center are the records of the National Jewish Hospital in Denver and the correspondence and papers of such physicians, scientists, and engineers as Henry H. Hart, Max Honigbaum, Myer E. Jaffa (see also Be.05), Samuel Lilienthal, and Adolph Sutro (see also Sa.02, St.02). The Blumenthal Library contains correspondence and papers for Walter C. Lowdermilk (see also Be.05), as well as archival materials pertaining to Albert Einstein. The Western Jewish History Center and Blumenthal Library have limited hours.

Be.05

The Bancroft Library
University of California
Berkeley, CA 94720
(415) 642-0959 (History of Science
and Technology Program)

The holdings of the Manuscripts Division at The Bancroft Library, including collections pertaining to science and technology, can be consulted in the Heller Reading Room at the Bancroft. It is necessary to make prior arrangements to use some of the collections described below, especially those for which no finding aid is available or those stored off campus.

Accademia del Cimento, Florence.
72/251. 1 box (17th century). Academy documents, including scientific papers of its secretary, Lorenzo Magalotti.

Adams, Frank (1875–1967), irrigation engineer and economist.
C-D4022. Oral history concerning irrigation, reclamation, and water administration.

Albertus, Carl.
74/207. 1 portfolio. Correspondence and papers relating to his work with Magnavox.

Alexander, Annie Montague (1867–1950), zoologist, UCB.
67/121. 1 box (1907–1949). Correspondence regarding establishment and work of the Museum of Vertebrate Zoology.

69/15. 1 v. Biographical sketch recounting her field trips, friendships with Louise Kellogg and Joseph Grinnell, and work at the Museum of Vertebrate Zoology.

C-B1003. 2 boxes, 3 v. (ca. 1904–1930), finding aid. Letters concerning the Museum of Vertebrate Zoology and a 1904 African safari; safari scrapbook, including photos.

Allen, Joel A. (1838–1921).

> 67/125. 1 portfolio (1883–1884), finding aid. Correspondence and papers, mainly relating to the founding of the American Ornithologists' Union.

Alvarez, Luis W. (1911–   ), professor of physics, UCB; Nobel laureate.

> 76/161. 7 cartons (ca. 1956–1970). Bubble chamber logs, including photographs.

> 84/82. 11 cartons (ca. 1956–1978), folder list. Three alphabetical series of correspondence and subject files, including materials on the Pyramid Project.

> Videotaped interview.

American Institute of Physics.

> Microfilm of transcripts of oral history interviews with Luis W. Alvarez, Felix Bloch, Ira Sprague Bowen, Lee Alvin DuBridge, William Alfred Fowler, George Gamow, Jesse Leonard Greenstein, Ivan Robert King, Robert Benjamin Leighton, Nicholas Ulrich Mayall, Frank Oppenheimer, Myron Spinrad, David Locke Webster, Clyde Wiegand, Robert R. Wilson, and Dean E. Wooldridge.

American Philosophical Society.

> 75/50, no. 3. 1 folder. List of manuscripts and papers given to the Society by the UCB Department of Genetics.

Amerine, Maynard A. (1911–   ), professor of viticulture and œnology, UCD.

> 73/53. Oral history, "The University of California and the state's wine industry."

Anger, Hal O. (1920–   ), biophysicist at Donner Laboratory, UCB.

> Oral history (HSTP), including discussion of the scintillation camera.

Appert, Kurt E. electrical engineer and entrepreneur.

> Oral history (HSTP), "Electrical engineering and the Lenkurt Electric Company."

Archive for History of Quantum Physics.
See Be.03.

Arnold, James R. (1923–), professor of chemistry, UCD.
84/16. Oral history (HSTP), "Nuclear chemistry at Princeton, Chicago, and San Diego." Includes discussions of his work with W. F. Libby and the research of Harold Urey and Stanley Miller.

Arnstein, Lawrence.
C-D4082. Oral history, "Community service in California public health and social welfare." See also Be.05 Warren.

Ashburner, William (1831–1887).
C-B504. 13 folders (1864–1882). Letters to William H. Brewer, some concerning the California Geological Survey; plus letters from Mrs. Ashburner.

Ashton, John Oliver.
79/114. 1 carton (ca. 1938–1959), folder list. Correspondence, articles, biographical materials, etc., relating to the history of radio and to Lee de Forest.

Astronomical Society of the Pacific.
78/124. 1 box (ca. 1909–1969). Organization records and correspondence.

Audubon Association.
C-A389. 3 boxes (1917–1945), finding aid. Correspondence, bird lists, minute book (1930–1944), and miscellaneous accounts (1926–1940).

Aughey, Samuel (1832–1912), Wyoming territorial geologist.
P-M30. 4 l. (1885). Dictation describing geological surveys and scientific writings.

Auhagen, Wilhelm.
71/147. 13 v. (ca. 1879–1883). Lecture notes taken while a student of astronomy in Germany.

Babbage, Charles (1792–1871).

73/160, no. 21. Letter (n.y.) to an unidentified correspondent.

Babcock, Ernest Brown (1877–1954), professor of genetics and department chairman, UCB.

81/139. 1 box (ca. 1909–1949). Reports, budget data, and correspondence regarding the College of Agriculture and his own teaching career.

Baldwin, John S.

Z-Z100:150. Letter (1833) to Julius Quesnel regarding fossils discovered in York, Canada.

Bandelier, Adolphe F. A. (1840–1914).

Film P-E210. 1453 exposures. Microfilm of typed transcript of his journals (1880–1885) and of pictorial materials assembled for a history of missions in New Mexico, Arizona, and Mexico, for the Jubilee of Pope Leo XIII. Includes materials concerning archæology.

Banks, Harvey O. (1910–   ), civil engineer.

Oral history, "California Water Project, 1955–1961."

Barlow, Chester (1873?–1902), California ornithologist.

76/24. 1 portfolio (1897–1899). Miscellaneous papers, including clippings, a letter from Joseph Grinnell, and postcards sent by Barlow.

Bascom, William R. (1912–1981), professor of anthropology and director of Lowie Museum, UCB.

80/1. 7 cartons, folder list. Correspondence and papers.

80/143. 3 cartons, folder list. Additions, primarily manuscripts of writings on African culture, with some related correspondence.

82/163. 46 cartons, folder list. Correspondence and papers regarding field work on African folklore and culture.

84/6. 4 cartons. Primarily field notes and early manuscripts about Yoruba social organization.

Beals, Ralph Leon (1901–), professor of anthropology, UCLA.

78/128. Oral history.

Beattie, Margaret (1893–1976), professor of public health, UCB.

77/45. 1 v. (ca. 1960–1970). Record of her career, including an account of her research on diagnosis of communicable diseases.

Martin Behaim's globe and Humboldt's essay.

M-M24. 47 l. (1852). Translation of F. W. Ghillany's preface to his *Geschichte des Seefahrers Ritter Martin Behaim* (Nürnberg, 1853) and of the essay by A. von Humboldt contained therein.

Bélidor, Bernard Forest de (1697?–1761), French mathematician and engineer.

73/2. 1 v. (ca. 1756). Manuscripts about the theory of compression and experiments with land mines.

Bernstein, Benjamin A. (1881–1964), professor of mathematics, UCB.

C-B969. 1 box, 3 cartons, 1 portfolio (1929–1963). Correspondence, manuscripts of articles, notes, and related papers concerning research in mathematical logic and teaching at Berkeley.

Berry, Samuel Stillman (1887–1984), zoologist.

81/169, no. 2. Oral history, primarily concerned with malacology.

Birge, Raymond Thayer (1887–1980), professor of physics and department chairman, UCB.

73/79. 41 boxes, 25 cartons, oversize materials (ca. 1909–1969), finding aid. Correspondence, manuscripts and reprints of his writings, research data and notebooks, speeches, course materials, correspondence and papers relating to university matters and professional organizations, etc. Includes his unpublished history (in 4 v.) of the Berkeley physics department.

82/29. 5 cartons (ca. 1914–1965), folder list. Additions, including subject files, course materials, correspondence, manuscripts of papers and talks, bibliography and departmental budget card files.

C-D4080. Oral history, including discussion of the Radiation Laboratory.

Bishop, Louis Bennett.

75/140. 34 boxes. Correspondence and papers, removed from the Cooper Ornithological Society collection.

Blaney, Harry French (1892–), irrigation engineer.

69/106. Oral history, including discussion of water conservation, agricultural issues, and soil science. See also Be.06.

The Blochmans of San Francisco, Santa Maria, and Berkeley.

73/122, no. 145. 19 l. Copy of typescript family history, by David F. Hoexter with Mary R. Hoexter. Discusses, among other topics, Ida Blochman's interest in botany.

Bohr, Niels (1885–1962), Nobel laureate.

80/89. 1 portfolio (1922). Notes on Bohr's Göttingen lectures on the semi-mechanical quantum theory of atomic structure.

Bolt, Richard Arthur (1880–1959).

69/111. 1 portfolio. Reminiscences concerning his career in public health.

C-D5165. 1 portfolio (ca. 1958). Typescript, "History of hygiene and public health at the University of California."

Bonar, Lee, and Lincoln Constance, professors of botany, UCB.

75/50, no. 11. 17 l. "Partial history of the Berkeley Department of Botany, 1890–1973." Includes information on botanical instruction in the College of Agriculture.

Boodt, William A.

C-R149. 165 p. Holograph manuscript of a book, "Practical drafting room mathematics" (1957).

Bookman, Max (1910–), consulting civil engineer.

69/104. Oral history, including discussion of water resources.

Boot, Henry A. H. (1917–), scientific officer, Royal Navy Scientific Service.

84/21. Oral history (HSTP), "The cavity magnetron and radar developments," discussing research at the University of Birmingham and the use of the magnetron in radar during World War II.

Born, James L. (1915–), director of medical operations at Donner Laboratory, UCB.

Oral history (HSTP).

Bošković, Rudjer Josip (1711–1787), Jesuit; professor of mathematics and physics, College of Rome; director of the Brera Observatory.

72/238. 15 boxes, 7 v., finding aid. Manuscripts and drafts of scientific papers and belles lettres, letters from several hundred correspondents, letters by Bošković, notebook and address book, biographical materials, earlier catalogs of the collection, plus microfilm and photocopies of documents in other repositories.

83/136. Photocopies of his letters (1756–1769) to Giovan Stefano Conti. Originals in the Brera Observatory archives.

Botta, Paul Emile (1802–1870), French naturalist.

70/97. 1 v. (1826–1829). Journal of zoological and ethnological observations made during a Pacific expedition.

Bowles, Edward (1897–), professor of electrical communications, MIT.

Oral history (HSTP).

Bowman, Karl M., psychiatrist; superintendent of Langley Porter Clinic.

70/79. Oral history.

Bracelin, Nina Floy.

68/132. 1 box, 4 card file boxes (1938–1963). Papers and research materials on food preservation; card files about botanical specimens; correspondence, some relating to Ynés Mexía; chemistry course materials.

82/143, no. 46. Oral history regarding the Mexía botanical collections.

Bradbury, Norris E.

See Be.05 Los Alamos Scientific Laboratory.

Bray, William Crowell (1879–1946), professor of chemistry, UCB.

73/55. 4 boxes (ca. 1890–1945), finding aid. Correspondence, drafts, articles, notes, and reprints.

C-Z151. Letter (1944) to Mel Gorman concerning a problem in chemistry.

Brewer, William Henry (1828–1910), botanist; assistant to Josiah D. Whitney on the California Geological Survey.

79/65. 1 box (1862–1909). Letters written to Brewer, plus photocopies of documents concerning the Survey.

C-B312. 5 boxes (1860–1884). Correspondence with Josiah D. Whitney, chiefly concerning the Survey.

C-B313–328. 16 v. (1861–1864). Field books with geological notes and observations.

C-B329–332. 4 v. (1860–1864). Diaries and journals recording the geological survey of California.

C-B333. 2 v. (1860–1864). Typed transcripts of letters from Brewer to his family, with a chronology of the geological survey of California and notes by Francis P. Farquhar. Originals in Yale Library.

C-B1069. 1 box (1860–1865). Records kept during the California Geological Survey, including Brewer's report to J. D. Whitney (1862), field notes, and accounts.

Bridges, Lyman, civil engineer.

78/28. 1 box. Correspondence and papers.

Brobeck, William M. (1908–), assistant director of the Radiation Laboratory, UCB; president of William M. Brobeck Associates.

84/14. Oral history (HSTP), "Engineering and big-machine physics at the Radiation Laboratory."

Brode, Bernice Bidwell (1901–).

76/187. 1 box. Memoir and photographs of Los Alamos laboratory during World War II.

75/50, no. 6. "In memoriam of Laura Fermi," recalling a dinner party in 1948.

Bryant, Harold Child (1886–), ornithologist.

80/10. 1 box. Notes on geographical distribution of game animals in California.

C-D4088. Oral history including discussion of conservation, natural history, and the National Park Service.

Burbank, Luther (1849–1926).

73/122, no. 13, 173. Letters (1902–1903), one with descriptions of the Shasta Daisy and Burbank's new hybrid clematis.

C-Z34. Two letters (1909) to Sam Lovett about the Wonderberry.

82/143, no. 20. Letter (1922) about his namesake.

Burke, Joseph.

Film P-W33. 114 exposures. Microfilm of letters (1843–1847) to Sir William Jackson Hooker, chiefly concerning botanical explorations in Canada and the American West. Originals at the Royal Botanic Garden, London.

69/63. Letter (1862) and obituary.

Butters, Charles (1854–1933).

C-B748. 1 carton (ca. 1894–1933). Scrapbooks, photographs, a few letters, and clippings concerning mining engineering and metallurgy.

C-D5048:10. Biography of Butters (part of the Ralph L. Phelps papers).

Buttner, Harold (1892–1979), vice-president, International Telephone and Telegraph.

84/13. Oral history (HSTP), "Research and development in radio and electronics, 1915–1974." Includes discussion of Bell Laboratories and Charles Litton.

Byerly, Perry (1897–1978), professor of seismology, UCB.

85/69. 6 cartons. Correspondence and papers, including materials relating to the UC network of seismographic stations.

C-Z171. 1 folder (1928). Correspondence with Arthur Brisbane concerning earthquake prediction.

Cajori, Florian (1859–1930), professor of history of mathematics, UCB.

C-B1006. 2 boxes, 5 cartons (1908–1928), folder list. Correspondence, notebooks, manuscripts of writings, notes, teaching materials, slides and photographs, etc. Mainly concerning the history of mathematics.

California. Department of Public Health. Bureau of Industrial Health.

C-A295. 1 box (1942–1946). Correspondence relating primarily to wartime nutrition programs.

———. Geological Survey.

C-B436. 4 letters concerning the Survey (1861–1866).

C-A183:1–22. 1 box, 332 maps (ca. 1844–1873), finding aid. Data, notes, maps, etc.

Calvin, Melvin (1911–), professor of chemistry, UCB; Nobel laureate.

Oral history (HSTP), "Chemistry and chemical biodynamics at Berkeley, 1937–1980." Includes discussions of the Manhattan Engineer District and Lawrence Berkeley Laboratory.

Cameron, Donald R. (1907–), professor of genetics, UCB.

76/70. 1 box, 1 carton, folder list. Correspondence, arranged alphabetically; subject files; annotated photographs; notebooks.

Camp, Charles Lewis (1893–), professor of paleontology, UCB.

73/178. 25 boxes, folder list. Correspondence and papers.

Phonotape 120. Oral history and related materials.

C-R143. Card file. Bibliography of Lorenzo Gordin Yates.

Campbell, Kenneth, son of William Wallace Campbell.

72/124. Oral history, "Life on Mount Hamilton, 1899-1913," with comments by C. Donald Shane concerning the history of Lick Observatory. See also SC.01.

Campbell, William Wallace (1862–1938), astronomer and university president, UCB.

71/219. 1 v. (ca. 1903–1928). Postcards, many from astronomers, to Campbell and to his family.

Canada. Geological Survey.

P-C37. Also available on microfilm. Memorandum on the published report of the Geological Survey of Canada (1871–1878).

Cavaciocchi, Michele.

76/2. 1 v. (1788). Notes on lectures by Vincenzo Mazzoni on spheres, geography, and astronomy.

Cerda, Tomás.

71/164. 2 v. Treatises (in Spanish) on statics, hydrostatics, and hydraulics.

Chadbourne, H. L.

81/169, no. 4. 62 l. (1982). Typescript, "William J. Clarke and the first American radio company."

Chamberlain, Owen (1920–), professor of physics, UCB; Nobel laureate.

Oral history (HSTP).

Chambers, Willie Lee (1878–1966).

67/131. 1 portfolio, 2 boxes, 1 carton (ca. 1891–1964). Correspondence, papers, and photographs; also some records of the Cooper Ornithological Club.

Chance, Ruth Clouse, director of the Rosenberg Foundation.

77/67. Oral history, "At the heart of grants for youth." Includes discussion of public health issues.

Chaney, Ralph Works (1890–1971), paleobotanist, UCB; conservationist.

C-D4073. Oral history.

Chevalier, François (?–1748).

76/3. 1 v. Textbook on mechanics.

Chickering, Allen Lawrence (1877–1958).

68/5. 1 v. (1923). Diary of a trip to Baja California, with descriptions of flora and fauna.

Chodorow, Marvin, and Charles Susskind.

75/50, no. 2. 1 v. (ca. 1960). "Fundamentals of microwave tubes."

Christy, Samuel Benedict (1853–1914), professor and dean of mining and metallurgy, UCB.

C-B1010. 7 boxes, 2 v. (ca. 1880–1905), folder list. Correspondence, subject files, scrapbooks, diagrams, photographs, ledgers, etc. Includes materials relating to the Department of Mining and Metallurgy.

Churchill, Alexander Lyman.

C-B811. 1 portfolio (1865–1908). Diary (1885–1886) kept while serving as ship's engineer; vouchers; and correspondence, some with the consulting engineer of the U.S. Revenue Marine.

Churchman, C. West (1913–   ), professor of business administration, UCB.

C-B1009. 1 box (ca. 1945–1947), folder list. Correspondence, manuscripts of writings, and research materials, mainly concerning statistics and operations research.

Clark, Nathan C.

78/33. Oral history, "Sierra Club leader, outdoorsman, and engineer." Includes discussion of vacuum tubes.

Clement, Lewis M. (1892–).

77/92. 1 box (ca. 1938–1975). Reminiscences, drafts of papers, correspondence, photographs, etc., relating to radio engineering and the history of commercial radio.

Coast Manufacturing and Supply Company

80/126. 2 boxes, 1 carton (ca. 1867–1939). Company records, including letterbooks, ledgers, and manuals, for the Oakland-based manufacturer of fuses.

Coffman, John Daniel (1882–1973), chief forester, National Park Service.

78/140. 1 box (1908–1952). Diaries describing work for the Park Service, association with the Forest Service, travels, etc.

74/69. Oral history, including discussion of fire control.

Communications Satellite Corporation (COMSAT).

85/97. 26 cartons (ca. 1963–1979). Records concerning the development of communications satellite systems and the establishment of COMSAT.

Cook, Sherburne Friend (1896–1974), professor of physiology, UCB.

80/35. 1 box, 6 cartons, 8 card file boxes, oversize materials, folder list. Research notes and data; transcripts of documents; notebooks and journals, including those for a trip with Carl Sauer; calculations and tables; maps and charts; card files with bibliographical information and notes on towns, many relating to Mexico and missions.

79/21. 8 cartons, 1 card file box, folder list. Research notes and data, reprints, manuscripts of writings, some correspondence, samples and slides of bone tissue, reports, etc. Includes materials on California Indians and on soil analysis.

Film C-A288. 12 reels. Microfilms of Bancroft Library and National Archives material pertaining to California Indians, assembled by Cook for presentation at U.S. Indian Claims Commission hearings.

Cooper, James Graham (1830–1902).

75/4. 1 portfolio (1853–1864). Letters to his family, some referring to his work for the Smithsonian Institution and the U.S. Geological Survey.

Cooper Ornithological Society.

74/144. 38 boxes, 8 cartons, folder list. Society records, field notes, etc.

82/7. 4 cartons. Additions.

Cotes, Roger (1682–1716).

72/12. 1 v. (ca. 1707). Lectures on hydrostatics and pneumatic experiments.

Crandall, Howard (1921–), chemist and specialist in operations research, Standard Oil of California.

Oral history (HSTP), including discussions of chemistry at UCB, the Materials Testing Accelerator project at the Radiation Laboratory, and operations research and computer development at Standard Oil.

Cremer, John (?–1822).

P-N134. Journal. Appended to the journal is an unexplained letter (1830) from Jedediah A. Smith concerning botanical specimens.

Crown Zellerbach.

80/71. Oral history series, "Timber, technology, and corporate development in the Pacific Northwest, 1920–1965."

Cruess, William Vere (1886–1968), professor of food technology, UCB.

68/101. Oral history, "A half century in food and wine technology."

Cutter Laboratories.

76/50. 2-volume oral history discussing pharmaceuticals and medicine, with supplementary documentary material. Covers the period 1897–1972.

Dakin, Frederick Holroyd.
73/92. 9 cartons, maps (ca. 1902–1966), folder list. Field notes, correspondence, subject files, maps, photographs, reports, claims, legal documents, etc., mainly relating to mining engineering.

Dana, Samuel T. (1883–1978), dean of the school of natural resources, University of Michigan.
68/104. Oral history, "The development of forestry."

Darwin, Charles Robert (1809–1882).
74/78. 1 portfolio (1878–1881). Correspondence, including letters to Asa Gray and to Benjamin Peirce.

Davidson, George (1825–1911), professor of geography, UCB.
C-B490. 64 boxes, 30 cartons, oversize material (ca. 1845–1911), finding aid. Correspondence, diaries, manuscripts, notebooks, drawings, subject files, photographs, and maps concerning the University of California, geography, navigation, commissions, and the U.S. Coast and Geodetic Survey. Includes family papers.

C-D788. 1 folder (ca. 1880), also available on microfilm. "A scientific life."

C-E131. 1 portfolio (1879). Autobiography and essay on irrigation.

74/176. Photocopies of 54 letters (1851–1860) to his mother-in-law.

Film 73/11. 3 reels. Microfilm of letterbooks (1846–1883). Originals at the California Academy of Sciences.

De scientiarvm & artivm vanitatem.
3Ms B53 D4. 1 v. (17—).

Deans, James (1827–?).
P-C38. 17 p. (1876). Papers relating to the cairns and shellmounds of Vancouver Island.

Demarcaciones y sondas.

Film 67/189. 1 reel. Microfilm copy of demarcations and soundings (1761) from Veracruz to Anegada, including sailing directions and related information. Originals in the Admiralty Library, London.

Dennes, William R. (1898–), professor of philosophy, UCB.

70/155. Oral history, "Philosophy and the University since 1915," including comments about Los Alamos and J. Robert Oppenheimer.

Dental history project.

See SF.02.

Derleth, Charles (1874–1956), dean of the College of Engineering, UCB.

C-B717. 2 boxes, 1 carton. Correspondence, reports, clippings, and photographs (1936–1938) relating to work as consulting engineer for the Golden Gate Bridge; scrapbooks concerning the 1906 San Francisco earthquake and fire. Other papers available at Be.06.

Dexter, Henry.

C-Y197. Letter (1875) to the California Academy of Sciences.

Dieter, Nannielou Hepburn (1926–), radioastronomer.

78/20. 1 box (ca. 1947–1974). Letters from Bart J. Bok, Cecilia Payne-Gaposchkin, and S. B. Pickelner; photographs; notes from a course given by Helen D. Prince.

Doble, Abner (1890–1961).

77/183. 2 boxes (ca. 1910–1954). Correspondence, patents, and design notebooks for Doble steam cars and locomotives.

Doble, Robert McE., consulting engineer.

70/61. 1 portfolio (1879–1910). Letters, reports, reprints, and maps relating to hydroelectric power and irrigation.

Doble papers.

C-B827. 6 boxes, 4 cartons (1912–1960), finding aid. Papers of Abner and Warren Doble relating to the Doble steam car and engine. Includes correspondence, diaries, notes, reports, photographs, and blueprints.

Dodd, Coleman.

75/50, no. 1. 1 v. (1945). Klystron survey, written for Sperry Gyroscope Company, describing tubes and customer comments.

Dollar collection.

69/113. 78 cartons, 72 v. (ca. 1872–1967), finding aid. Family papers, plus business records of divisions of the Robert Dollar Company, including the Globe Wireless Company.

Drawings of all the coasts.

Film 67/188. Microfilm of drawings of Brazil (early 17th century) in the Admiralty Library, London.

Drury, Aubrey (1891–1959).

67/152. 1 box (1931–1938). Correspondence, promotional material, and clippings, relating in part to the need for a large reflector at Lick Observatory.

73/3. 5 cartons, 8 card file boxes, folder list. Correspondence and papers, including correspondence regarding adoption of the metric system in the United States. Also 81/52.

Durbin, Patricia W. (1927–), physiologist, UCB.

75/50, no. 15. 15 p. (1976). Typed transcript of a speech, "Reminiscences of Crocker Laboratory," covering its establishment and work during the 1930s and World War II, with special mention of Joseph G. Hamilton.

Eastwood, John Samuel.

69/11. Notebook of excerpts from publications concerning Eastwood's career as a civil-hydraulic engineer. Assembled by Marguerite Eastwood Welch.

Eddy Tree Breeding Station.

75/33. Oral history series, including interviews with Gladys

(Mrs. Lloyd) Austin, Francis I. Righter, William G. Cumming, Alfred R. Liddicote, Jack Carpender, and Nicholas T. Mirov. Also the Forest History series at Be.05.

Edmonston, Robert M. (1925–).

76/45. Oral history, "Preliminary studies: The California Water Plan."

Einarsson, Sturla (1879–), professor of astronomy, UCB.

75/50, no. 8. Notebook (ca. 1906) concerning astronomy and photography, kept while Einarsson was a graduate student and assistant at Berkeley.

Einstein, Albert (1879–1955), Nobel laureate.

75/50, no. 17. Letter from Carl-Ferdinand Universität (Prague) to Einstein (1911). Calculations on verso in Einstein's hand and that of another, possibly Mileva Einstein.

Eitel-McCullough, Inc.

77/110. 4 cartons (ca. 1934–1965), folder list. Company records; some correspondence. See also Be.05 McCullough.

Elliott, George Henry (1831–1900).

C-B302. 1 v. (1858–1860). Correspondence while on duty with the U.S. Coast Survey.

Ellis, N. Randall.

C-B443. 10 cartons, oversize material (ca. 1908–1925), folder list. Correspondence, notes, articles, legal briefs, notebooks, reports, bound volumes of exhibits and statistics, blueprints, etc. Primarily concerning Ellis' work with Pacific Gas and Electric and with the city of San Francisco as a valuation engineer.

Emerson, William Otto (1855?–1940).

C-R48. 5 folders. Collection concerning natural history.

Essig, Edward Oliver (1884–), professor of entomology, UCB.

74/30. 3 cartons, folder list. Primarily diaries, scrapbooks, and mimeographed material relating to the history of entomology; some correspondence.

Euclides.
2Ms QA451 E8. Manuscript book (ca. 1460) of *Elements*.

Evans, Griffith C. (1887–1973), professor of mathematics and department chairman, UCB.

74/178. 24 cartons, partial folder list. Correspondence, drafts and reprints of writings, subject files, course materials, etc., concerning the Department of Mathematics, his own teaching and research, the Applied Mathematics Panel during World War II, and professional organizations. Includes some correspondence and papers of his father and his son.

Evans, Herbert McLean (1882–1971), professor of biology, UCB; historian of science.

75/126. 18 boxes, 58 cartons, folder list. Correspondence, subject files, notes and drafts of articles and lectures, course materials, research notes, reprints, records of clinical studies, notebooks, personal papers, photographs. Includes research material on vitamin E, estrus cycles, history of science, etc.

84/42. 2 boxes (ca. 1926–1964), folder list. Correspondence with Miriam Simpson; reprints, lectures, bibliography, reminiscences, and other materials concerning Evans' work in physiology and history of science.

84/122. 20 cartons, carton list. Correspondence, bibliography and biography cards, reprints, and clippings, etc. Primarily concerned with book collecting and research in the history of science.

71/132. Tape-recorded recollections of John Muir, by Evans and others.

Everson, William [Brother Antoninus], poet.

75/5. 16 boxes, 13 cartons, 5 oversize boxes, finding aid. Correspondence and maps, most concerning his literary career; some materials related to astrology.

Exploratorium.

84/43. 1 box (ca. 1968–1982). Records and correspondence of Frank Oppenheimer regarding the founding and operations of the Exploratorium, a participatory science museum in San Francisco.

Farquhar, Francis P. (1887–).

Film C-B542, 543. Microfilm of correspondence and materials relating to Clarence King, some concerning the 40th Parallel Exploration, and of observations and accounts of the ascent of Mount Whitney (1873). Originals of the former in private hands; of the latter at Harvard.

Fava, Florence M.

78/11. 1 portfolio (1971–1973). Copies of transcripts regarding archæological findings at an Indian site in Los Altos Hills.

Fernán Núñez collection.

225 v. Primarily 16th- and 17th-century Spanish and Portuguese manuscripts, including some scientific treatises.

Fidler, Harold Alvin (1910–).

74/198. 1 carton (1962–1972), folder list. Records relating to the Atomic Energy Commission Labor-Management Advisory Committee, including minutes of meetings.

Field, Cyrus West (1819–1892).

Ms TK5611 F45. 2 letters (1859) about financing the Atlantic cable.

Fischer, Emil (1852–1919), German organic chemist; Nobel laureate.

71/95. 37 boxes, 12 cartons, oversize material (1876–1919), finding aid. Correspondence, manuscripts and reprints of writings, research files, laboratory notebooks (including those of students), subject files, materials concerning work during World War I; plus condolence letters to Hermann Fischer and Hermann Fischer's students' notebooks. Includes materials concerning the Kaiser Wilhelm Gesellschaft.

Fischer, Otto.

83/43. 2 cartons, oversize materials. Documents relating to the Union Diesel Engine Company.

Fisher, Gerhard (1899–), engineer, Federal Telegraph Company; founder of Fisher Research Laboratory.

83/139. Oral history (HSTP), "Radio engineering, 1923–1967."

Foster, Herbert Bismarck (1885–1968), university engineer, UC.

C-D4072. Oral history, "The role of the Engineer's Office in the development of the University of California campuses."

Fox, Denis Llewellyn (1901–), professor of marine biochemistry, Scripps Institution.

76/172. 260 l. Typescript autobiography, "Again the scene," including letters.

Franceschi, Francesco (1843–?).

70/11. 20 boxes (ca. 1904–1913), finding aid. Correspondence, manuscripts of writings, and lists relating to work in acclimatizing plants. Also including some later correspondence (1913–1918).

Fritz, Emanuel (1886–), professor of forestry, UCB.

C-B728. 6 cartons, 1 box, finding aid. Correspondence, subject files, circulars, clippings, reports, reprints, photographs, etc. Relating primarily to national and state forests, fires and controlled burning, Save-the-Redwoods League, and legislation.

68/20, 78/42. 21 boxes, 34 cartons, 11 card file boxes, folder list. Additions.

C-Z61. "Preserving the history of the forest" (1956).

74/20. Oral history, "Teacher, editor, and forestry consultant."

Froman, Darol K.

See Be.05 Los Alamos Scientific Laboratory.

Fuller, Leonard Franklin (1890–), professor of electrical engineering, UCB; vice-president of Federal Telegraph Company.

79/91. 1 carton (ca. 1915–1958), folder list. Data and

reports on radio equipment, Federal Telegraph, and the Stanford High Voltage Laboratory; patent litigation materials; photographs; professional journals; subject files. Includes materials relating to the Radiation Laboratory.

77/105. Oral history (HSTP), with supplementary volume of documents.

Galilei, Galileo (1564–1642).

f3Ms QB36 G26. 19 v. Copies (18th-century) of 17th-century correspondence, for an edition planned by Giovanni Battista Clemente Nelli.

Galloway, John Debo (1869–1943).

67/40. 1 box (1920–1940), finding aid. Correspondence, clippings, and promotional material, some concerning raising funds for Lick Observatory.

Gardner, Max William (1890–1979), professor of plant pathology, UCB.

80/60. 1 box, 2 cartons (ca. 1910–1968), partial folder list. Experiment station and seminar materials, correspondence, photographs, histories of the Department of Plant Pathology, clippings, minutes of meetings, manuscripts of writings, etc.

Gaytes, Herbert (1871–1965), engineer.

67/71. 15 cartons, 4 scrapbooks, folder list. Correspondence of Gaytes and his family; diaries; notebooks, field notes, and project files, many concerning water power projects; scrapbooks about World War I; corporate reports and records; maps and blueprints, etc.

George, Thomas C.

C-E184. History (1884) of the University of the Pacific, with a description of its scientific equipment.

Gerbode, Frank (1907–   ), cardiovascular surgeon.

Oral history (HSTP).

Giauque, William Francis (1895–1982), professor of chemistry, UCB; Nobel laureate.

83/37. 59 cartons, oversize materials (ca. 1930–1981), partial folder list. Correspondence, subject files, research data, notebooks, grant proposals and reports, course materials, drafts and reprints of writings, etc. Includes materials relating to the Low Temperature Laboratory and research projects headed by Giauque.

Gifford, Edward Winslow (1887–1959).

C-B768. 1 portfolio (1916–1955). Mainly correspondence with Edward Sapir concerning kinship terms and American Indian linguistics. Includes letters from Herbert L. Mason and others.

Gilfillan, S. Colum (1889–   ), sociologist.

75/50. Transcript of interview (1976) concerning methods for predicting technological innovations and trends.

Gillespie, Chester G. (1884–1970).

83/25. Oral history, "Origins and early years of the [California] Bureau of Sanitary Engineering."

Gilman, Daniel Coit (1831–1908).

C-B522. Portfolio (1871–1875). Correspondence, mainly relating to the geological survey of California and to the University of California.

Gofman, John (1918–   ), professor of medical physics, UCB.

Oral history (HSTP), including discussions of Donner Laboratory and Lawrence Livermore Laboratory.

Goldhaber, Sulamith, (1923–1965).

79/25. 1 box (ca. 1955–1957). Papers relating to the first antiproton annihilation event.

Goldschmidt, Albert.

C-E135–137. 3 v. (1873). "Cartography of the Pacific coast of North America and of the eastern coasts of Mexico and Central America," prepared for H. H. Bancroft.

Goldschmidt, Richard Benedict (1878–1958), professor of zoology (genetics), UCB.

    72/241. 5 boxes, 4 cartons (1900–1956), finding aid. Correspondence, some about work in Germany; research notes and data; lectures, etc. Includes correspondence of Curt Stern about Goldschmidt.

Goodspeed, Thomas Harper (1887–1966).

    80/138. 3 boxes (1926–1950), finding aid. Mainly copies of letters by Goodspeed and by Helen-Mar Wheeler relating to research on tobacco.

    Z-D133. 542 p. (1961?), also available on microfilm. Printer's copy for the 2nd edition of *Plant hunters in the Andes*.

Gray, Asa (1810–1888), botanist.

    C-B468, pt. 1 (in the Bidwell papers). 28 letters (1877–1886) to Mr. and Mrs. John Bidwell, concerning California botany.

Grayson, Andrew Jackson (1819–1869).

    C-B514. 23 boxes, 2 reels microfilm, finding aid. Correspondence, including correspondence of Grayson's wife; diary; field notebooks; manuscripts of his writings; watercolors and photographs; clippings; microfilm copies of materials elsewhere. Primarily relating to ornithological studies.

Gregory, John (1724–1773).

    73/95. Four-volume Scottish text on medicine, each volume indexed.

Grendon, Alexander (1899–), biophysicist at Donner Laboratory, UCB.

    Oral history (HSTP), "Research with Hardin Jones at Donner Laboratory, 1957–1978."

Griffith, Benjamin G.

    75/146. 1 box (1924–1927), also available on microfilm. Letters written to his family while working as a radio operator in Alaska; snapshots; miscellaneous materials.

Grinnell Naturalists Society

68/27. 2 boxes, card file (1940–1952), finding aid. Records of the organization founded by students and staff of the Museum of Vertebrate Zoology, UCB; photographs.

Grinnell, Joseph (1877–1939).

C-B995. 21 boxes (ca. 1884–1938), finding aid. Correspondence and papers relating to Cooper Ornithological Club and to zoological specimens from Alaska and California.

73/25. 5 cartons (ca. 1893–1939), folder list. Notebooks and field notes; correspondence, including letters of the Cooper Ornithological Club; manuscripts of writings; clippings; course notes, etc. Includes correspondence and papers of Hilda Wood Grinnell.

Guerlac, Henry F.

Film 74/205. Microfilm of "The history of radar in World War II" (2 v.). Original in the National Archives.

Gulick, John Thomas (1841–1923).

72/161. 4 v. (1841–1916). Correspondence and journals, some relating to conchology and evolution, mining, and travels in the Pacific. Typed transcripts.

Hacke, William.

See Be.05 South Sea Waggoner.

Hale, George Ellery (1868–1938), director of the Mount Wilson Observatory.

Film 73/86. 100 reels, finding aid. Microfilm edition of correspondence and papers (1882–1937), including materials concerning California Institute of Technology, the National Research Council, and the National Academy of Sciences. Originals at the library of the Mount Wilson and Palomar Observatories.

Hall, Harvey Monroe (1874–1932).

C-B908. 3 boxes, 1 carton (1896–1932), finding aid. Correspondence, notes, reprints, clippings, photographs, etc., relating to his career in botany and to his family.

Hamilton, Joseph Gilbert (1907–1957), director of Crocker Laboratory, UCB.

> 77/66. 24 v. (ca. 1936–1941). Notebooks on nuclear medicine experiments at the Radiation Laboratory.

Hansen, William Webster (1909–1949), professor of physics, Stanford University.

> 75/104. 4 v. (1941–1944). Mimeograph notes for lectures on microwaves given at the MIT Radiation Laboratory, edited by S. Seely and E. C. Pollard.

Hassid, William Zev (1897–1974), professor of biochemistry, UCB.

> 79/32. 5 boxes (ca. 1915–1974), finding aid. Correspondence, reprints, manuscripts of articles and talks, etc., relating to research in carbohydrate chemistry at UCB.

Hedgpeth, Joel Walker (1911–   ), director of the Pacific Marine Station, Bodega.

> 78/156. 3 boxes, 4 v. (ca. 1958–1978), finding aid. Correspondence concerning Pacific Gas and Electric's proposed nuclear plant; scrapbooks of clippings; articles and related papers, with information on the effects of radioactivity on marine life; and materials concerning the Bodega Marine Laboratory.

Heintz, Ralph Morell (1892–1980), inventor and entrepreneur; co-founder of Jack and Heintz, Inc.

> 77/175. 9 cartons, 2 scrapbooks (ca. 1920–1976), folder list. Correspondence, documents, scrapbooks, and photographs relating to Heintz's inventions and business ventures.

> 75/50, no. 14. Interview transcript, including discussion of vacuum tube development in the Bay Area, invention of the gammatron, Heintz's companies, and the Robert Dollar Company.

> 84/20. Oral history (HSTP), "Technical innovation and business in the Bay Area."

Heizer, Robert Fleming (1915–1979), professor of anthropology (archæology), UCB.

> 67/179. 2 boxes (1962–1967), finding aid. Materials relating to the discovery of the Anza burial site in Mexico and to the publication of Heizer's book on the subject.
>
> C-I11. 1 portfolio. Correspondence, notes, photographs, diagrams, maps, etc., concerning the archæological investigation of Sutter's mill in Coloma.
>
> C-R142. 1 portfolio (1923–1931). Clippings concerning Indians of southern California.
>
> 77/71. 1 box, 3 cartons. Papers relating to his editorship of the California volume of the North American Indian handbook.
>
> 77/72. 1 carton. Correspondence and papers concerning his research on Cabrillo's grave marker.
>
> 78/17, 80/11. 41 cartons, 2 boxes, oversize materials, folder list. Correspondence, subject files, research notes, manuscripts of writings, lecture notes, photographs, maps, etc., concerning California Indians, archæological investigations, and other topics.
>
> 82/41. 2 cartons. Correspondence and papers, mainly concerning research on California Indians and their land claims.

Helmholz, A. Carl (1915–), professor of physics, UCB.

> Oral history (HSTP).

Henderson, Malcolm (1904–1975), professor of physics, Catholic University of America.

> 84/150. 1 box (ca. 1935–1970). Correspondence and papers.
>
> Oral history, including discussion of work at the Radiation Laboratory, UCB.

Herschel, Sir John Frederick William, bart. (1792–1871), British astronomer.

> Film 73/165. 32 reels, reproduction and publication restricted. Microfilm of correspondence, manuscripts, diaries, works, journals. Originals at the University of Texas, Austin, library.

Hildebrand, Joel Henry (1881–1983), professor of chemistry, UCB.

71/69. 1 box (ca. 1936–1979). Correspondence, manuscripts of writings, articles, lecture notes, bibliography, etc. Includes materials about the Sierra Club.

85/68. 6 cartons. Research notes, correspondence, etc.

C-D4052. Oral history, "Chemistry, education, and the University of California" (1960).

75/26. Oral history.

81/136. Oral history (HSTP), "Physical chemistry at the University of California, Berkeley" (1974–1978).

Hildebrand, Milton (1918–).

Z-C221. Letters (1941–1942) to his family, describing a UC expedition to El Salvador to collect specimens of animals, plants, and fossils.

Hilgard family, especially Eugene W. Hilgard, professor of soil science, UCB.

C-B972. 32 boxes, 7 cartons, slides, oversize materials, 31 v. (1848–1945), finding aid. Correspondence, manuscripts, subject files, lecture notes, etc., concerning soil science and the Northern Transcontinental Survey. Includes papers of family members, some concerning the U.S. Coast and Geodetic Survey.

Hill, Charles Barton (1863–1910).

C-B358. 1 portfolio, 1 v. (1881–1910). Manuscripts, articles, notes, and clippings concerning astronomy and Lick Observatory.

Hinds, Julian (1881–), dam engineer.

71/268. Oral history.

Hodge, Frederick Webb (1864–1956), head of the Bureau of American Ethnology.

C-D4016. Oral history, including discussion of the Bureau and the U.S. Geological Survey.

Hoffmann, Charles Frederick,

C-B505. 1 box (1862–1910). 121 letters to Josiah D. Whitney, chiefly concerning cartography for the California Geological Survey; notes; 11 personal letters to William H. Brewer.

Hohl, L. J.

68/92. 3 cartons, maps (ca. 1908–1931), folder list. Correspondence, maps, and reports, primarily concerning work as a consulting mining engineer.

Holmes, Jack A. (1911–1967), professor of educational psychology, UCB.

68/155. 22 cartons, folder list. Correspondence; research notes, data, and reports; course materials; manuscripts and reprints of writings; materials concerning campus and professional activities; other subject files.

Holmes, Samuel Jackson (1868–1964).

C-B935. 2 boxes, 6 cartons, oversize materials, folder list. Correspondence, manuscripts and reprints of writings, notes, clippings, personal papers, etc. Includes materials about zoology, evolution, and eugenics, and about the loyalty oath.

Howard, Walter Lafayette (1872–1949).

C-R6. 1 carton. Papers relating to his biography of Luther Burbank, including correspondence and records of interviews with persons who knew Burbank, abstracts from Burbank's scrapbooks, and sale literature.

Huffman, Eugene, analytical chemist, Lawrence Berkeley Laboratory.

74/91. 1 box, 2 cartons (ca. 1945–1967), folder list. Correspondence, subject files, and reprints.

International Union of Pure and Applied Physics.

78/162. 1 carton (1966), folder list. Correspondence and papers concerned with the Berkeley conference on High-Energy Physics.

Isaac, Leo (1892–1970), researcher, U.S. Forest Service.

70/167. Oral history, "Douglas fir research in the Pacific Northwest." Includes discussion of forest genetics.

Jaffa, Myer E. (1857–1931), professor of agricultural chemistry and nutrition, UCB; director of the State Food and Drug Laboratory.

C-B1013. 1 carton, 1 oversize portfolio (ca. 1896–1923), folder list. Subject files, correspondence, and clippings concerning research on agricultural chemistry. See also Be.04.

Jaffé, George Cecil (1880–1965), professor of physics, University of Giessen and Louisiana State University.

76/210. 2 boxes (ca. 1902–1962), finding aid. Correspondence, notes, drafts, and documents concerning his research and teaching in Germany and the United States. Includes materials relating to his paper on Ludwig Boltzmann.

James, Edwin (1797–1861).

Film P-W12. 220 exposures (1820–1827). Microfilm of diary kept during Long's expedition to the Rockies; of a journal kept while serving as army surgeon; of miscellaneous geological, botanical, and other scientific notes. Originals at Columbia University Library.

Jenkins, Francis Arthur (1899–1960), professor of physics, UCB.

78/54. 2 cartons, folder list. Correspondence and subject files, chiefly concerned with university committees and professional organizations.

Jenkins, James Angus (1904–1965), professor of genetics, UCB.

C-B996. 19 cartons (1930–1965), finding aid. Correspondence, course materials, reports, minutes, manuscripts and reprints of papers, notebooks, research notes, etc., including materials relating to his research on tomato plants.

Jenkins, Katharine.

83/117. 13 cartons, folder list. Research materials on cochineal, Mexican textiles, etc., including notebooks, subject files, and file boxes. Also 70/202.

Jensen, Dilworth Darwin (1910–), entomologist.

74/28. 13 cartons, folder list. Correspondence and papers, including files concerning research proposals and grants, notebooks, etc.

Jepson, Willis Linn (1867–1946), professor of botany, UCB.

75/50, no. 10. 1 folder. Biography by L. Constance.

C-B437. 10 folders. Letters to Ernest Dawson and Charles Yale (1924–1940), from Jepson and nine others. The holdings of the Jepson Herbarium on the Berkeley campus also include some of Jepson's correspondence and papers.

Jones, Guy E.

C-R114. 1 portfolio (ca. 1850–1923). Materials collected relating to medicine and public health in California. Includes materials about the 1850 cholera epidemic, notes on Thomas M. Logan, a diary, clippings, reprints, statistics, and charts.

Jones, Hardin Blair (1914–1978), professor of medical physics and physiology, UCB.

79/112. 15 boxes, 5 cartons (ca. 1937–1978), finding aid. Correspondence; drafts and reprints; research notes on medical physics, radiation, drug abuse, and other medical subjects; articles about Jones. Also photographs, film, and phonotapes. Includes materials concerning Donner Laboratory.

Jordan, David Starr (1851–1931), university president, Stanford.

77/47. 1 portfolio (1900–1924). Letters and papers.

Jordan, Peter (1751–1825).

76/124. 2 v. *Vorlesungen* on agriculture and related scientific subjects (Vienna, 1805).

Joslyn, Maynard A. (1904–), professor of food science and technology, UCB.

75/51. Oral history, "A technologist views the California wine industry."

Judah, Theodore Dehone (1826–1862).

C-B575. 1 box (1851–1862). Family papers, including a herbarium of California wildflowers.

Kaiser Permanente, health maintenance organization.
Oral history series (in progress).

Kamen, Martin David (1913–), professor of biochemistry, UC San Diego.

76/16. 2 cartons (ca. 1937–1945), folder list. Research notes and notebooks, correspondence, and manuscripts of articles, chiefly concerning research by Kamen and Samuel Ruben on carbon-14 and photosynthesis.

80/108. Oral history (HSTP), "Physics, politics, and photosynthesis."

Kimble, George W.

68/76. 1 box (1872–1910). Scrapbook, notes, index of minerals in his collection, geological reports, and survey field notes.

Kingsley, John Sterling (1854–1929).

C-D5057. 266 l. (ca. 1927). Reminiscences, emphasizing American zoology in the late 19th century.

Kittel, Charles (1916–), professor of physics, UCB.

83/140. 40 l., restricted. Responses to a questionnaire from the American Institute of Physics, relating to his research in solid-state physics and work with U.S. Naval Ordnance, MIT, Bell Laboratories, and UCB.

Knudsen, Vern Oliver (1893–1974), professor of physics, UCLA.

75/11. Oral history, including discussion of Bell Laboratories, acoustics, and submarine detection during World War II.

Kofoid, Charles Atwood (1865–1947), professor of zoology, UCB.

67/126. 1 portfolio (1921–1922), finding aid. Correspondence about nomination as professor of protozoology, School of Tropic Medicine, Bombay.

82/39. 1 portfolio (1916). Correspondence concerning hunting in India, mainly received while on a scientific expedition.

Koshland, Daniel Edward (1920–), professor of biochemistry, UCB.

84/33. 5 cartons. Correspondence and papers.

Kotcher, Ezra (1903–), aeronautical engineer.

Oral history (HSTP), including discussion of innovations in aeronautics.

Kroeber, Alfred Louis (1876–1960), professor of anthropology, UCB.

78/22. 3 cartons, folder list. Research notes, manuscripts of writings, correspondence, descriptions of photographed artifacts, articles, maps, many concerning California Indians and Indian Land Claims hearings. Includes material relating to Kroeber's death, e.g., an interview with Claude Lévi-Strauss.

83/109. 1 box. Letters to B. Pinner (restricted), translations of German verse, typed transcripts of his Gimbel lectures, and materials about insectivores and about metrics.

C-B925. 32 boxes, 12 cartons (ca. 1900–1960), finding aid. Correspondence, manuscripts, field notes on linguistics, notes, subject files, etc., concerning anthropology, especially his work with California Indians.

71/83. 2 boxes, 12 cartons, 6 card file drawers, 1 v. (ca. 1905–1959), finding aid. Additions.

Kroeber-Quinn, Theodora (1897–1979), anthropologist.

69/145. 1 box, 2 cartons, oversize materials. Correspondence, including replies to letters sent to Alfred Kroeber.

83/27. Oral history, "Timeless woman, writer, and interpreter of the California Indian world."

Krueger, Albert P. (1908–1982), professor of bacteriology, UCB.

84/7. 23 cartons (ca. 1960–1980), folder list. Subject files, reports, notes, correspondence, project proposals, research data, course materials, graphs, photographs, etc. Much material concerning his research on air ions.

Krueger, Myron E. (1890–), professor of forestry, UCB.

Oral history.

Kurti, Nicholas (1908–), professor of physics, Oxford University.

Oral history (HSTP), including discussion of low-temperature physics. Restricted.

Lammerts, Walter E.

84/149. 3 cartons (ca. 1960–80). Correspondence relating to the Creation Research Society.

Langelier, Wilfred F. (1886–), civil engineer, UCB.

83/26. Oral history, "Teaching, research, and consultation in water purification and sewage treatment, University of California at Berkeley, 1916–1955."

Langmuir, Irving (1881–1957), physicist and chemist, General Electric Research Laboratory.

Film 80/128. Microfilm copy of correspondence (1914–1921) about radio, high-power vacuum tubes, and the theories of G. N. Lewis. Originals in the Library of Congress.

Lawrence, Ernest Orlando (1901–1958), professor of physics, UCB; Nobel laureate.

72/117. 2 boxes, 48 cartons (ca. 1920–1968), finding aid; some materials restricted. Correspondence, manuscripts of articles and speeches, notebooks, diagrams, photographs, research notes, etc.; administrative records for the Radiation Laboratory; some papers of Joseph Hamilton and Donald Cooksey. Includes records of the Manhattan Engineer District and Atomic Energy Commission.

84/129. Additions.

Lawson, Andrew Cowper (1861–1952), geologist.

C-B602. 19 boxes, 14 cartons, 1 portfolio (ca. 1885–1951), finding aid. Correspondence, notebooks, manuscripts and reprints of articles, lectures, poems, maps, photographs, business papers, etc. Includes materials concerning the Canadian and U.S. Geological Surveys and seismology.

Lawson, James S. (1828–?).

P-A44. 142 l. (1879). Autobiography, describing his experiences with George Davidson and the U.S. Coast Survey (1850–1878).

LeConte family papers.

C-B1014, 70/24, C-B452. 2 cartons, 7 boxes, 3 portfolios, finding aid. Correspondence, notes, and writings of John LeConte (1818–1891), professor of physics and university president, UC; Joseph LeConte (1828–1901), professor of geology, UC; and others.

LeConte, John and Joseph.

C-D769. 1 portfolio. Biographical sketches of John and Joseph LeConte, in the hand of Alfred Bates.

LeConte, John Lawrence (1825–1883).

C-Y283. Petition to the California legislature (ca. March 1850), concerning the need for a geologic survey of the state.

LeConte, Joseph (1828–1901).

81/49. 1 box (1858–1900). Mainly incoming correspondence.

Lee, John.

76/98. 1 portfolio (1858–1866). Minutes of the Balloon Committee of the British Association for the Advancement of Science, and correspondence concerning the use of balloons for scientific research.

Legge, Roy A. (1895–   ).

73/122, no. 67. Typescript, "The Mount Whitney Stone Hut" (1975), concerning a Smithsonian project in 1909 to measure solar energy.

Lennette, Edwin, virologist, California Department of Health and the School of Public Health, UCB.

Oral history (HSTP).

Lenzen, Victor Fritz (1890–1975), professor of physics, UCB.

76/206. 5 boxes, 6 cartons, 1 oversize folder (ca. 1904–1975), finding aid. Correspondence, manuscripts, course materials, research and lecture notes, reprints, administrative and personal papers; also an article written and annotated by Albert Einstein. Includes materials concerning philosophy and archæology.

C-D4090. Oral history, "Physics and philosophy."

Leopold, Aldo Starker (1913–), professor of zoology, UCB.

81/61. 8 cartons, 1 oversize notebook (ca. 1925–1972), partial folder list. Correspondence, subject files, notebooks, reports, course materials, photographs and maps, journals, manuscripts of books and articles, etc., concerning forestry and wildlife surveys and management.

85/98. 5 cartons. Subject files, manuscripts of writings, etc.

Lesquereux, Leo (1806–1889).

C-B563. 58 letters (1862–1878) to Henry N. Bolander, concerning mosses and other plants in California.

Leuschner, Armin Otto (1868–1953), professor of astronomy, UCB.

C-B1016. 5 boxes, 4 cartons, oversize material (ca. 1875–1951), finding aid. Correspondence; manuscripts of writings; personal papers; subject files, mainly concerning institutions and organizations; bibliography cards; class work and observation notebooks; photographs; printed materials.

Levi ben Gershon (1288–1344).

2Ms B485 L3A7. 112 l. (14th–15th century). Supercommentary on Averroes on Aristotle, *De generatione et corruptione, Meteorologica*, etc.

Lewis, Gilbert Newton (1875–1946), professor of chemistry, UCB.

73/70. 1 carton. Correspondence and papers.

Lewis, Rubin M. (1899–1976), thoracic surgeon.

80/63. Oral history, "From Butte to Berkeley."

Libby, Willard F. (1908–1980), professor of chemistry, UCB and UCLA; Nobel laureate.

83/120. Oral history discussing such topics as the Manhattan Engineer District, carbon–14 research, and the Atomic Energy Commission.

Lippincott, Donald Knudsen (1889–).

79/39. 6 cartons (ca. 1913–1956), carton list. Papers related to his service in the U.S. Signal Corps and work as an attorney specializing in radio and television patents. Includes materials concerning Philo T. Farnsworth.

Litton, Charles Vincent (1904–1973), engineer and entrepreneur.

75/5. 13 boxes, 6 cartons, oversize material (ca. 1912–1972), finding aid. Correspondence, notes and notebooks, patents, drawings, reports, etc., concerning vacuum tubes, Litton's businesses, and theories of physics.

Livingood, John J. (1903–), physicist, Argonne National Laboratory.

84/8. Oral history (HSTP), "Big-machine physics at Berkeley, Harvard, and Argonne." Includes discussion of the Radio Research Laboratory.

Loeb, Leonard Benedict (1891–1978), professor of physics, UCB.

73/6. 23 boxes (ca. 1916–1970), finding aid. Correspondence, notes, papers and speeches, reports by Loeb and his students, clippings, etc., concerning research on gaseous discharge physics, teaching activities, and work on spark plugs during World War I.

79/105. 2 boxes, 31 cartons, folder list. Correspondence; subject files; course and department materials; travel files; manuscripts of books, articles, and speeches; reprints; referee reports; notebooks on gas ion research; materials concerning the U.S. Naval Reserve.

C-D5184, 76/89. Autobiography prepared for the American Institute of Physics (1962), plus a list of Loeb's publications and those of his students. A supplement covers 1962–1975. Chiefly concerns his work in physics and association with the U.S. Navy.

Lofgren, Edward J. (1914–), physicist and associate director, Lawrence Berkeley Laboratory.

Oral history (HSTP).

Lombard, Henri Clermond (1805–1895).

76/4. 1 v. (ca. 1820). Notes on a zoology course given by A. P. de Candolle in Geneva.

Los Alamos Scientific Laboratory.

81/75. Oral history interviews with Norris E. Bradbury, John H. Manley, Darol K. Froman, J. Carson Mark, and Raemer E. Schreiber concerning wartime and postwar research and connections between Los Alamos and the Atomic Energy Commission, UCB, and Lawrence Livermore Laboratory. See also Be.05 Underhill.

Louderback, George Davis (1874–1957), professor of geology and dean of the College of Letters and Sciences, UCB.

83/65. 2 boxes, 4 cartons, oversize materials (ca. 1914–1916), finding aid. Correspondence, field notes, maps, journals, photographs, accounts, and other materials concerning his geological expedition to China for Standard Oil Company of New York.

C-B928. 32 boxes, 10 cartons, 1 portfolio (ca. 1900–1956), finding aid. Correspondence, diaries, and papers.

Lowdermilk, Walter Clay (1888–1974), forester and hydrologist.

72/206. 5 cartons, 1 tube (ca. 1912–1969). Correspondence, diaries, field notes, manuscripts of his articles, reports, photographs, and certificates.

69/36. 3 cartons, 1 portfolio (1914–1961), finding aid. Materials relating to surveys of agricultural land use, including correspondence, logs, photographs, and reports; manuscripts and reprints of his articles on conservation, forestry, and land reclamation.

70/3. Oral history, "Soil, forest, and water conservation and reclamation in China, Israel, Africa, and the United States."

Lowie, Robert Harry (1883–1957), professor of anthropology, UCB.

C-B927. 14 boxes, 7 cartons, 1 portfolio, 1 card file (ca. 1893–1957), finding aid. Correspondence, notebooks and photographs for his research on American Indians and Indian linguistics; lecture notes; diaries; manuscripts of his writings; subject files, etc.

80/9. 2 boxes, 1 card file box. Additions.

Maindron, Ernest (1838–1908).

73/187. "Mélanges académiques" (1885–1896), including his publications on the history of science, especially in France.

Majors, Harry M.

70/92. 1 box, also available on microfilm. "Science and exploration on the northwest coast of North America, 1542–1841" (completed in 1969).

77/99. 1205 l. Critical bibliography.

Maker, Frank L., petroleum engineer.

76/202. 2 cartons, folder list. Course notes, correspondence, reprints, subject files, design notes, etc. Concerning the design of particle accelerators, petroleum engineering, and patents.

Manhattan Project: Official history and documents.

12 reels of microfilm, finding aid. Originals in the National Archives.

Manley, John.

See Be.05 Los Alamos Scientific Laboratory.

Manson, Marsden (1850–1931).

C-B416. 2 cartons (1887–1927), finding aid. Correspondence, reports, manuscripts, printed materials, clippings, maps, pictures, etc., chiefly concerning San Francisco's water supply and his work as city engineer (1908–1912).

Marg, Elwin (1918–), professor of optometry, UCB.

85/71. 2 cartons. Correspondence and papers relating to his research and to development of optometric instruments.

Mark, J. Carson.

See Be.05 Los Alamos Scientific Laboratory.

Marmont, George H., professor of engineering, Naval Postgraduate School, Monterey.

5 cartons. Correspondence and papers relating to electronics and its applications in medicine and physiology.

Marshall, Lauriston Calvert (1902–1979), physicist, professor of electrical engineering, and director of the microwave power laboratory, UCB.

80/127. 15 cartons (ca. 1924–1979), folder list. Correspondence, research notes, course materials, reprints, etc., including materials relating to accelerator research, space sciences, and electronics.

80/141. 1 box, 2 cartons. Additions.

Marshall, Robert Bradford (1867–1949).

C-B511. 23 boxes, 4 scrapbooks, 1 portfolio, 1 volume, finding aid. Correspondence, notes, manuscripts of writings, speeches, memoranda, clippings, and scrapbooks, mainly concerning the Marshall Plan for water development, conservation, Hetch-Hetchy, and national parks.

Martens, W. L. F. (?–1931).

74/183. 1 box (ca. 1868–1926). Correspondence, papers relating to patented improvements in surveying instruments, and miscellaneous notes and accounts. Includes letters concerning Lick Observatory and the California Academy of Sciences.

Massachusetts Institute of Technology. Radiation Laboratory.

82/92. 17 cartons (ca. 1939–1946). Laboratory technical reports and card index to reports.

Mathias, Mildred Esther (1906–), botanist, UCB and UCLA.

> 83/18. Oral history, "Among the plants of the earth." See also Be.05 Rowntree.

Matthes, François Emile (1874–1948), topographer and geologist, U.S. Geological Survey.

> C-B821. 3 cartons, 5 boxes, 1 oversize folder (ca. 1896–1948), finding aid. Correspondence, manuscripts and reprints of writings, diaries, notes, clippings, etc. Includes papers of his wife.

May, Bernice H.

> 76/175. Oral history, including discussion of the wartime science training program at UCB.

McCarthy, Edmund Patrick (1832–?).

> C-D822. 50 l. Dictation concerning emigration to California (1855–1856) and work as a hydraulic engineer and miner. Recorded for H. H. Bancroft.

McCulloch, Walter (1905–1973), dean of the Forestry School, Oregon State University.

> 69/30. Oral history, "Forestry and education in Oregon, 1937–1966." Includes discussion of conservation issues.

McCullough, Jack, co-founder of Eitel-McCullough.

> Oral history (HSTP), including discussion of the klystron, Heintz and Kaufman, Varian, and Charles Litton.

McGaughey, Percy H. (1904–1975), director of the Sanitary Engineering Research Laboratory, UCB.

> 75/25. Oral history, "The Sanitary Engineering Research Laboratory, 1950–1972."

Mclaughlin, Donald H. (1891–), professor of mining engineering and geology, UCB.

> 76/63. Oral history, "Careers in mining geology and management, university governance and teaching." Includes discussion of the U.S. Geological Survey and Homestake Mining Company.

McMillan, Edwin Mattison (1907–), professor of physics, UCB; Nobel laureate.

74/151. 26 cartons, folder list. Synchrotron notebooks, files, research data, diagrams and plans, etc.

81/134. Oral history for the American Institute of Physics, including discussion of the development of radar and the work of the UCB Radiation Laboratory and the Manhattan Engineer District.

80/129. Diary (1927) kept during a prize tour in Europe.

McMillan, Elsie Walford (Blumer).

78/38. 1 v. (ca. 1973). "The atom and Eve," reminiscences of Los Alamos and the Radiation Laboratory.

Mead, Elwood (1859–1936), professor of engineering, UCB.

C-B1041. 22 cartons, folder list. Correspondence, research data, manuscripts and reprints of writings, photographs and slides, hearing transcripts, students' theses, etc., relating in part to agriculture, irrigation, and engineering.

Medical College of Pennsylvania.

Oral history series concerning women in medicine.

Merriam, Clinton Hart (1855–1942), naturalist, chief of the U.S. Biological Survey.

C-B520. 2 cartons (1874–ca. 1938), partial folder list. Correspondence; notes and articles, many concerning American Indians; government documents and hearings; scientific miscellany; clippings.

C-R21. 1 carton. California place names file.

72/56. 1 carton. Clippings and notes concerning anthropology, archæology, Indians of North America, Eskimos, Maya culture, prehistoric man, arctic expeditions, etc.

79/138. 5 boxes, 16 cartons, 1 portfolio. Correspondence; extensive subject files, especially about bears; manuscripts and reprints of his writings, etc.

80/18. 26 cartons, card file boxes. Data concerning California and other American Indians, copies of documents, clippings, photographs, ethnographic information, data on Indian welfare, manuscripts of writings, maps, and correspondence.

82/45, 83/129. 2 boxes, 26 cartons. Additions.

Merriam, John Campbell (1869–1945), professor of paleontology, UCB.

C-B970. 15 boxes, 1 carton (ca. 1904–1934), finding aid. Correspondence, manuscripts and reprints of articles, and miscellaneous papers.

71/100. 6 boxes, 5 cartons (ca. 1911–1943), folder list. Correspondence, manuscripts of writings and speeches, subject files, including materials relating to La Brea pits and conservation.

Metcalf, Woodbridge (1888–1972), forester, UC Agricultural Extension Service.

C-B1018. 1 box, 1 portfolio (1927–1932). Correspondence and papers concerning fire protection and prevention in California.

73/180. 24 cartons, folder list. Correspondence, reports, subject files, field notes, pamphlets and reprints, diaries, photographs and slides, manuscripts of writings, botanical specimens, notebooks, etc., primarily relating to trees in California.

77/95. 1 box, 8 cartons, folder list. Correspondence, notes, course materials, data, reports, published materials, maps, manuscripts of writings, notebooks, note cards, etc., much of it concerned with forest fires and research on eucalyptus.

84/118. 2 cartons. Files relating to the Society of American Foresters, Northern California Section.

69/150. Oral history, "Extension forester, 1926–1956."

Mexía family papers.

M-B1. 5 boxes, 2 cartons, portfolios (1694–1951). Correspondence and papers, including some papers of Ynés Mexía.

Mexía, Ynés (1870–1938), botanist, Smithsonian Institution and UCB.

68/130. 11 boxes, 2 cartons, 1 v. (1910–1938). Correspondence, biographical and family information, and notebooks and file boxes concerning botanical collections. Some materials assembled by N. F. Bracelin.

Mexico. Archivo General de la Nacíon. Epidemias.

9 reels of microfilm concerning 18th- and 19th-century medicine.

Meyer, Karl F. (1884–1974), professor of bacteriology, UCB; director of the Hooper Foundation, UCSF.

76/42. 5 boxes, 108 cartons, folder list. Correspondence, research notes, diaries, course materials and lecture notes, reports, films, manuscripts and reprints of writings, etc.

83/45. 17 cartons, folder list. Additions.

76/138. Oral history, "Medical research and public health." Includes discussion of bacteriology and pathology.

Miller, Alden H. (1906–1965).

67/3. 2 boxes, 2 card file boxes (ca. 1949–1965). Papers relating to the Cooper Ornithological Society.

68/82. 11 cartons. Additions.

Miller, Herman Potts, Jr., radio engineer.

78/58. 8 cartons (ca. 1914–1967), folder list. Correspondence, manuscripts of writings, designs, test results, photographs, published materials, notebooks, manuals, proposals, reports, etc. Includes materials relating to tubes, antennas, transmitters, Federal Telegraph Company, and Stanford University.

Miller, Loye Holmes (1874–1970), zoologist and ornithologist, UCLA.

C-B997. 1 box (1899–1957), finding aid. Correspondence and diaries, relating in part to investigations of fossil beds.

67/209. 1 box (1910–1966). Letters, clippings, and photographs.

Phonotape 64. Tape-recorded interview.

71/87. Oral history, "The interpretive naturalist." Includes discussion of zoology and paleontology.

Miller, Stanley, (1930–), professor of chemistry, UC San Diego.

Oral history (HSTP).

Minkowski, Rudolph Leo B. (1895–1976), professor of astronomy, UCB.

81/34. 16 cartons (ca. 1902–1960), carton list. Research notes, photographs, correspondence, reprints, etc., concerning work in astronomy and astrophysics.

Mirov, Nicholas Tiho (1893–), plant physiologist, U.S. Forest Service.

67/198. 2 v. (1917–1963), restricted. Diary concerning emigration to America, research on pines, teaching geography at UCB; photographs; clippings, etc.

69/136, 72/18. 2 cartons. Correspondence and papers.

75/33. Oral history.

Mitchell, Marion O.

C-R95. 1 portfolio. Correspondence, clippings, and notes concerning Alexis von Schmidt (1821–1906), civil engineer.

Moore, Norman H. (1918–), and **Roy E. Woenne** (1916–).

84/17. Oral history (HSTP), "Vacuum tube and magnetron development."

Morgan, Agnes Fay (1884–1968), professor of home economics, UCB.

75/63. 11 boxes (ca. 1905–1966), folder list. Correspondence, drafts, reprints, manuscripts of articles and speeches, etc., concerning home economics, nutrition, and biochemistry.

Morris, Fred Ludwig (1873–1955).

C-B744. 10 cartons, maps (1894–1933), finding aid. Correspondence, maps, and papers concerning mines, mining operations, and petroleum engineering.

Morse, A. P., professor of mathematics, UCB.

84/134. 8 cartons. Correspondence, subject files, course materials, manuscripts and reprints of writings, etc.

Moyer, Burton Jones (1912–1973), professor of physics, UCB; dean of the College of Liberal Arts, University of Oregon.

74/89. 5 cartons, carton list. Correspondence, diaries, research notebooks, manuscripts and reprints of his writings, photographs, Ph.D. thesis, reports, biographical information, etc. Includes materials related to work on radiation protection at the Radiation Laboratory, UCB.

Mrak, Emil Marcel (1901–), professor of food technology, UCB; chancellor, UCD.

76/1. Oral history.

Muir, John (1838–1914), naturalist.

C-B468, pt. 1 (part of the Bidwell papers). 31 letters (1877–1912) to Mr. and Mrs. John Bidwell, chiefly concerning California plants, travels in California and Alaska, and glaciers.

C-H101. 1 box, 1 portfolio (ca. 1860–1914). Correspondence, reports, notes, and drawings concerning conservation, geology, and inventions.

70/17. 1 box (1909–1913). Letters to Katharine Putnam Hooker, plus other correspondence and articles.

Myers, William G. (1908–), professor of biophysics, Ohio State University.

Oral history (HSTP) concerning the early history of nuclear medicine.

National Research Council.

75/50, no. 4. "An experimental study of the quartz source-project of the supersound" by the San Pedro Anti-submarine Group (Throop College of Technology, 1919). Group members were John A. Anderson, H. D. Babcock, and Harris J. Ryan.

Needham, Paul R., professor of zoology, UCB.

67/59. 11 cartons, folder list. Subject files, research notes, correspondence, grant proposals, etc., mainly concerning research on fish and fisheries.

Nelson, Dewitt (1901–), professor of forestry, UCB.

77/117. Oral history, "Management of natural resources in California, 1925–1966." Includes discussion of the U.S. Forest Service and a supplementary volume of documentary material.

Newton, Sir Isaac (1642–1727).

75/101. "The accompts of Mr. Ambrose Warren" [1701]. Formerly in the Portsmouth collection of books and papers written by or belonging to Newton (Catalogue entry no. 308).

BS1556 M67 P5 1681 copy 2. Henry More, *A plain and continued exposition of the several prophecies or divine visions of the prophet Daniel* (London, 1681). With autograph and marginal notes by Newton.

Neyman, Jerzy (1894–1981), professor of statistics, UCB.

84/30. 60 cartons, folder list. Correspondence and subject files; manuscripts and reprints of his writings; travel files; research proposals, data, and reports; materials relating to the Statistical Laboratory and to war work, etc.

84/51. 5 cartons. Additions.

Northwestern Pacific Railway.

77/143. 1 carton (ca. 1914–1939), folder list. Diagrams, photographs, instruction books, and other manuscript materials concerning railway electrification.

Norton, Charles Eliot (1827–1908).

M-M1828. 1 portfolio (1879–1892). Letters to Henry Heynes concerning the Archæological Institute of America and the work of Adolph Bandelier in Mexico.

Olmo, Harold P. (1909–).

77/40. Oral history, "Plant genetics and new grape varieties."

Packard, Walter Eugene (1884–1966), agricultural engineer; superintendent of the UC Experimental Farm.

67/81. 10 boxes, 12 cartons, oversize portfolio (1899–1966), finding aid. Correspondence, manuscripts and copies of writings and speeches, reports, notes, photographs, and clippings; family papers.

71/16. Oral history.

Palache, Charles (1869–1954), curator of the Harvard Mineralogical Museum.

C-D5193. 27 l. Photocopy of his autobiography, including an account of his geological studies with Joseph LeConte and A. C. Lawson and his work in Germany and at Harvard.

C-F154. 2 folders. Diary of a Sierra trip with LeConte and the class of 1891; typed transcripts.

Palmer, Harold King (1878–?).

C-B712. 1 box (1903–1905). Letters to his family, describing his experiences on a UC expedition to Chile to build and operate the San Cristóbal observatory.

Pardies, Ignace Gaston (1636–1673).

3Ms QA33 P3. 109 f. Manuscripts of his *Elementa geometriae* and *Statica* (Padua, 1730).

Paris. Some account of the establishments [for] physical sciences.

72/86. Record of educational and scientific institutions, (ca. 1809), including professors' names, courses, and associated museums and libraries.

Parry, Charles Christopher (1823–1890).

C-B468, pt. 1 (part of the Bidwell papers). Letters concerning California botany.

Pauly, ——— , geographical engineer to the French king.

M-M1713. 14 l. History of the voyage of the abbé Chappe to California to observe the transit of Venus in 1769. Translation from the French.

Peirce, Charles Sanders (1839–1914).

69/12. 137 l. (ca. 1889). Photocopy of a report on gravity at four U.S. locations. Original in the archives of the U.S. Coast and Geodetic Survey.

Perham, Douglas, etc.

75/50, no. 12. 9 l. (1973). Description by Arthur L. Norberg of the papers of Perham, of Harold F. Elliott, and of the Royden Thornberg collection at the Foothill Electronics Museum.

Geology class field trip near Carmel (Thanksgiving 1891). Photograph by Andrew C. Lawson. University Archives, The Bancroft Library.

Perry, Newel Lewis (1873–1961).

67/33. 1 carton, 2 portfolios (1899–1961). Correspondence and papers, mainly concerning his work with the California Council for the Blind, together with notes, drafts, and reprints of his thesis in mathematics.

C-D4028. Oral history. See also C-D4027 at Be.05.

Pinart, Alphonse Louis (1852–1911).

Z-Z17. 24 v. (1870–1885), finding aid. Correspondence, notes, drafts and copies of articles, diaries, drawings and maps, many concerning Pinart's travels and research on linguistics and ethnology.

Additional materials on linguistics, ethnography, geography, and American Indians can be found in C-C62, M-M478–494, P-C40, P-C47, P-D17, P-K49, P-N13–41, Z-C8, and 68/51.

Pinger, Roland W.

C-B973. 15 v. (1923–1956). Notes and articles compiled while teaching in the UCB College of Engineering; ROTC instruction policy books.

Pollard, James F.

75/93. 2 v. (1912–1913). Thesis about the Oakland-Sacramento railway, written with C. C. Snyder for the UCB Department of Electrical Engineering; Pacific Gas and Electric notes and technical diary.

Poniatoff, Alexander (1892–1980), president of Ampex Corporation.

81/95. 1 box (ca. 1948–1964), partial folder list. Correspondence and brochures, including materials concerning magnetic tape recording. Additions include research notebooks (primarily photocopies of articles about medicine).

Oral history (HSTP), including discussion of his experiences in Russia, the founding of Ampex, and tape recording technology.

Popenoe, Edwin A.

75/82. 1 portfolio (1873–1878). Letters received, including those from John LeConte and George Horn concerning entomological specimens.

Porter, Robert Langley (1870–1965), dean of the Medical School, UC[SF].

C-D4076. Oral history, "Physician, teacher, and guardian of public health."

Powell, Richard Cheadle, professor of engineering, UCB.

75/28. 3 cartons, folder list. Subject fields, reprints, correspondence, notebooks, etc., including materials concerning power and light companies in Hawaii and elsewhere.

Powers, Stephen (1840–1904).

68/169. 1 portfolio (1875–1876). Photocopies of transcripts of letters to John Wesley Powell concerning Powers' manuscripts on California Indians. Originals in National Archives.

C-E162. 1 portfolio (ca. 1874). Notes on Indians and Indian languages. Also C-E83:8.

Pratt, Haraden (1891–1969).

72/116. 4 boxes (ca. 1908–1969), finding aid. Correspondence, notes, articles, and photographs concerning his early work in radio and research on the history of radio. Includes information about Federal Telegraph Company and other firms.

Proyecto de un puerto...

Z-D110. 1 portfolio (ca. 1900–1910). Copies of correspondence and reports, drawings, etc., presumably from the files of Arturo Castaño, engineer, concerning the proposed port on Bamborombón Bay, Buenos Aires.

Purcell, Edward Mills (1912–   ).

81/169, no. 3. Early typescript version of volume 2, "Electricity and magnetism," in the Berkeley physics course (1963). Includes Purcell's sketches and annotations by Purcell and Walter D. Knight.

Putnam, John Alpheus (1907–), professor of engineering, UCB.

67/10. 8 v. (1943–1958). Correspondence, writings, and proceedings of faculty seminars sponsored by Standard Oil of California.

RCA Short Wave Radio Station, Bolinas.

77/207. 1 carton, 3 rolls (ca. 1927–1962). Log books, reports, data, correspondence, photographs, blueprints, legal documents, etc.

Rabuel, [Claude] (1669–1728).

74/126. Treatises in Latin on practical geometry and mechanics (Lyon?, 1715).

Randall, Henry Irwin (1863–1937), engineer for Southern Pacific Railroad; rancher.

69/56. 9 cartons (ca. 1889–1936). Correspondence and reports, some concerning his engineering work; engineering notebooks.

Ransom, Leander (1800–1874).

C-B562. 1 portfolio. Correspondence, reminiscences, clippings, and photographs, etc., concerning Ransom, his work with the U.S. Engineer's Office, and his association with the California Academy of Sciences. Includes field notes relating to the establishment of the Mount Diablo base and meridian by Ransom in 1851. Assembled by Glenn B. Ashcroft.

Rawn, A. M. (1888–1968), sanitary engineer.

69/103. Oral history, including discussion of water resources and pollution.

Regnault, [Henri Victor] (1810–1878).

78/142. 1 v. "Traité[s] des logarithmes, des fractions, de la sphère, de la géometrie."

Reynolds, Wallace B. (1904–), business manager and managing engineer at the Radiation Laboratory, UCB.

80/61. 1 portfolio, scrapbooks (ca. 1931–1968). Correspondence, scrapbooks, and miscellaneous papers concerning work at

Oak Ridge during World War II and at the Radiation Laboratory.

84/9. Oral history (HSTP), "Laboratory management at Berkeley and Livermore." Includes discussion of the Manhattan Engineer District.

Rhiem, Johann Christoph.

75/50, no. 16. 34 l. Draft of a translation into French of Rhiem's *Dissertatio mathematica, de praestantia arithmeticae binariae prae decimali*... (Jena, 1718).

Rhodin, Carl J.

C-B1048. 1 portfolio (ca. 1931). Correspondence, mainly concerning proposed rapid transit for the Bay Area, with related reports and memoranda.

Richter, Clemens M. (1848–?).

71/57. 102 l. (1922). Autobiography and reminiscences, including discussion of medical practice and epidemiology.

Rising, Willard Bradley (1839–1910), professor of chemistry, UCB.

C-B1023. 11 boxes, 2 cartons (ca. 1866–1908), folder list. Correspondence, writings and speeches, research notes, department and course materials, notebooks, clippings, personal papers, etc.

Ritter, William Emerson (1856–1944), professor of zoology, UCB; director of Scripps Institution.

71/3. 23 boxes, 33 cartons (ca. 1879–1944). Correspondence, manuscripts and reprints of writings, field notes and drawings, annotated works by others, and clippings, primarily concerning marine biology and the establishment of Scripps Institution and of the Science Service.

Rix, Edward Austin (1855–1929), San Francisco engineer and inventor.

79/77. 1 box (ca. 1852–1933). Correspondence, notes, patents for pumps and drills, reminiscences, family letters, etc.

Roberts, Eugene L.
P-F358. 242 l. (1947). Account of the life of Benjamin Cluff (1858–?), including experiences on a scientific expedition to Mexico, Central America, and Colombia.

Rolle, Michel (1652–1719).
74/66. Transcript of his printed work on algebra (Paris, 1691).

Rowntree, Lester (1879–1979), horticulturalist.
80/8. Oral history, including discussion of botany and the California Academy of Sciences.

Ruben, Samuel.
See Be.05 Kamen.

Sauer, Carl O., professor of geography, UCB.
77/170. 21 boxes, 7 cartons, folder list and correspondent index. Correspondence, drafts and notes relating to publications and speeches, unpublished manuscripts, notebooks, photographs, research and field notes, course materials, etc.

Sauer, Martin.
P-K68. Also available on microfilm. Digest (1802) of an account of Billings' second astronomical and geographical expedition to northern Russia (1791–1792).

Schlesinger Library (Radcliffe College).
80/93. Oral history series about women in the birth control and maternal health movements.

Schneider, Albert (1863–1968).
C-B500. 1 box, 1 carton (ca. 1900–1920). Biographical information, miscellaneous writings, and photographs, some concerning his work in pharmacy and botany.

Schreiber, Raemer.
See Be.05 Los Alamos Scientific Laboratory.

Schulz, Helmut W. (1912–).

75/50, no. 9. Photocopy of an unpublished report, "Separation of the uranium isotopes by centrifugation," submitted to Harold Urey in 1940; plus biographical materials about Schulz and a copy of his letter to Urey.

Schumacher, Paul.

C-E187:1. Notes and drawings (1875) concerning ancient graves of California.

Scofield, Philip Forest (1901–), radio engineer.

75/50, no. 13. Report and transcript of interviews about his radio work and about Ralph Heintz.

Scott, Flora (1891–), professor of botany, UCLA.

74/124. Oral history.

Scott, Kenneth Gordon (1909–), director of the Radioactivity Research Center, UCSF.

80/81. 1 carton (ca. 1932–1974), folder list. Correspondence, drafts and reprints of writings, clippings, etc.

Oral history (HSTP), including discussion of early radioisotope research at Crocker Laboratory, UCB.

Seaborg, Glenn Theodore (1912–), professor of chemistry and chancellor, UCB; chairman of the Atomic Energy Commission; Nobel laureate.

Oral history (HSTP), "Nuclear research and national science policy."

Sefer mathematikah yashan w'nadir.

t4Ms QA99 S34. 178 p.

Segrè, Emilio (1905–), professor of physics, UCB; Nobel laureate.

78/72. 2 cartons (ca. 1955–1975). Manuscript of his biography of Enrico Fermi; course materials; manuscripts and photographs for *Nuclei and particles* (2nd ed.); correspondence and documents concerning the Piccioni court case.

Seismological Society of America.

77/164. 3 cartons (1911–1973), carton list. Society correspondence, arranged chronologically, including correspondence with George Louderback, Herbert Hoover, Perry Byerly, and Andrew Lawson; Society records and reports.

Senn, Milton J. E.

80/92. Set of four oral history interviews with leaders in the child guidance and clinic movement, including discussions of work at UC.

Setchell, William Albert (1864–1943).

C-B1034. 1 box, 2 cartons (1874–1937). Scrapbooks, photographs (transferred to Pictorial Collection), diary, and some correspondence, primarily concerning the University of California and botany.

Shane, Charles Donald and Mary Lea.

70/144–145. Oral history interviews about Lick Observatory and Donald Shane's work in astronomy. See also SC.01.

Shumate, C. Albert (1904–).

82/94. Oral history, "San Francisco physician, historian, and Catholic layman."

Sierra Club.

250+ cartons (ca. 1910–). Organization records and correspondence, including materials from the San Francisco (national) and Washington offices.

Silver, Samuel (1915–1976), professor of electrical engineering, UCB.

77/114. 20 boxes, 17 cartons (ca. 1939–1976), finding aid. Correspondence, manuscripts and reprints of writings, project proposals, course materials, research notes, administrative materials, photographs, etc., concerning engineering, electronics, and the UCB Space Sciences Laboratory.

Simon, Emil Jacob (1888–1963), radio engineer.

72/115. 4 boxes (ca. 1888–1963), finding aid. Autobiography with letters, photographs, and clippings documenting his work in radio; Lee de Forest memorabilia.

Sinclair, Donald Bellamy (1910–), engineer, General Radio Company.

84/19. Oral history (HSTP), "Radio engineering and research, 1926–1974," including discussion of the Radio Research Laboratory.

Siri, William Emil (1919–), biophysicist at Donner Laboratory, UCB.

80/4. Oral history, "Reflections on the Sierra Club, the environment, and mountaineering, 1950s–1970s." Includes discussion of the Manhattan Engineer District.

Oral history (HSTP), focusing on nuclear medicine at UCB.

Six Companies, Inc.

77/195. 6 v. (1931–1941). Minute books, scrapbooks, and other corporate records for the construction engineering firm (involving W., S., and K. Bechtel and Henry J. Kaiser). Includes materials relating to the construction of Boulder Dam.

Sizer, Frank L. (1856–1942).

67/15. 56 v. (1887–1942). Diaries, mainly concerning his experiences as a mining engineer.

Slate, Frederick (1852–1930), professor of physics, UCB.

73/136. 1 portfolio (1879–1894). Mainly correspondence with European firms about purchasing equipment for the Berkeley physics department.

C-B1051. 2 boxes, 1 portfolio (1862–1937), finding aid. Correspondence, diaries, manuscripts of speeches and articles (including a biography of Hilgard), diplomas, photographs, and family papers.

Smathers, James Field, inventor of an electric typewriter.

84/133. 1 box (1957–1961). Correspondence and other

materials assembled by Anne Sherman concerning Smathers' invention and IBM.

Smith, Charles Edward (1904–1967), dean of the School of Public Health, UCB.

73/49. 2 cartons, oversize materials (ca. 1938–1967). Reprints, biographical information, correspondence, engagement books, awards and photographs, etc.

Smith, Cyril Stanley (1903–), director of the Institute for the Study of Metals, University of Chicago; member of the General Advisory Committee, Atomic Energy Commission; professor of metallurgy, MIT.

81/137. Oral history (HSTP), "Metallurgy and atomic energy policy." Includes discussion of the Manhattan Engineer District.

Smith, Jeremiah Hobart.

C-G251. Notebook (ca. 1885) on mechanical engineering and drafting.

Smith, Ralph (1874–?), professor of plant pathology, UCB.

C-Z170. Typescript, "The beginning of modern plant pathology in California" (1953).

South Sea Waggoner.

M-M224. Also available on microfilm. Appendix to a South Sea Waggoner, probably made in the workshop of William Hack[e] (ca. 1685), based upon a Spanish *derretero* captured by Bartholomew Sharpe in 1680.

Film 67/147. Microfilm of the Huntington Library copy.

Film 68/37. 2 reels. Microfilm of originals in the John Carter Brown Library.

Film 71/114. 3 reels. Microfilm of William Hacke's manuscript atlases, "The Great South Sea of America" (1682–1698). Originals in the British Library.

Sowerby, Miss.

76/5. 2 v. Hand-colored drawings with brief text, based on Isaac Lea's *American conchology* (1839).

Sparkman, Philip Stedman (?–1907).

C-B1068. 1 portfolio (1896–1907). Correspondence with A. L. Kroeber, John Wesley Powell, and others; reminiscences of New Mexico in the 1880s; notes relating to San Diego county and genealogy. Includes materials concerning the Archæological Institute of America, the U.S. Bureau of American Ethnology, and the California Academy of Sciences.

Spier, Anna Hardwick (Gayton) (1899–1977), anthropologist, UCB.

79/6. 1 carton, 1 card file box (ca. 1925–1965), finding aid. Correspondence, field notes, manuscripts of her writings, and notes, primarily concerning the Yokuts Indians in California.

Spier, Leslie (1893–1961), anthropologist.

79/7. 2 boxes, 1 carton, 2 card file boxes (ca. 1924–1961), finding aid. Correspondence, notes, field notebooks, etc., primarily relating to his research on American Indians.

Spier, Robert F. G. (1922–).

79/5. 1 v. (1949–1950). Field notes concerning the Yokuts Indians.

Spieth, Herman Theodore (1905–), professor of zoology and chancellor, UCD.

80/110. Oral history, including discussion of entomology and evolution.

Spink, Henry Makinson.

71/136. Master's thesis in geography, "Economic development of the Pacific Coast of North America" (Liverpool, 1921).

Spira, Robert, mathematician.

84/62. 1 carton. Correspondence, writings, etc.

Stanley, Wendell Meredith (1904–1971), professor of biochemistry and molecular biology, director of the Virus Laboratory, UCB; Nobel laureate.

78/18. 1 box, 29 cartons (ca. 1930–1971), carton list. Correspondence, subject files, manuscripts of writings and speeches, reprints, course and departmental materials, photographs and negatives, etc.

84/28. 2 cartons. Additions.

Statewide Conference on Undergrounding Utilities.

75/50, no. 5. Papers presented at the Conference, held in Los Angeles (1967).

Stern, Otto (1888–1969), professor of physics, University of Hamburg and Carnegie Institute of Technology; Nobel laureate.

85/96. 6 cartons (ca. 1910–1968). Correspondence, notes, manuscripts and reprints of writings, photographs and awards, etc.

Stewart, George W. (1857–1931).

79/63. 1 box (ca. 1905–1930). Correspondence, including letters from Alfred Kroeber and C. Hart Merriam.

Stone, Andrew Jackson (1859–1918), collector, American Museum of Natural History.

P-K230. 1 carton (1896–1918). Diaries, drafts of books, copies of his reports to the Museum, and other papers.

Stone, Herbert (1883–?), geologist and engineer.

75/100. Memoirs (1970), including an account of his career and of his work with Lee de Forest.

Stone, Robert Spencer (1895–1966), professor of radiology, UCSF.

80/80. 1 box (ca. 1940–1956). Correspondence, manuscripts of writings, and notes.

Storer, Tracy I., and Ruth R. Storer.

77/49. A dual memoir and oral history concerning his work in zoology and hers in medicine. Includes discussions of UCD and of the Museum of Vertebrate Zoology at Berkeley.

Stratton, George Malcolm (1865–1967), professor of psychology, UCB.

C-B1032. 2 boxes, 8 cartons (ca. 1911–1956), finding aid. Correspondence; clippings; notes; manuscripts of lectures, books, and articles. Includes students' recollections in 1919 of the 1906 earthquake and fire in San Francisco.

71/155. 1 carton (ca. 1898–1957). Additions.

Struve, Otto (1897–1963), professor of astronomy, UCB.

67/135. 1 portfolio (ca. 1917–1966), finding aid. Correspondence, most of it prior to his appointment at Berkeley; bio-bibliographical materials; photographs and family papers. Includes materials concerning the International Astronomical Union.

81/35. 3 cartons (ca. 1949–1956), folder list. Mainly correspondence concerning professional organizations and meetings.

Sturtevant, William C.

77/60. 1 portfolio (1973–1975). Photocopies of correspondence with Robert Heizer and José L. Morales concerning Indians of the American West.

Sullivan, Celestine J.

C-B401. 3 letters (1918–1923) in the Chester Rowell papers, written while secretary of the League for the Conservation of Public Health.

Tarski, Alfred (1902–1983), professor of mathematics and logic, UCB.

84/69. 12 cartons, folder list. Correspondence, manuscripts of articles and books, preprints and reprints, unpublished writings, subject files, photographs and certificates, notebooks, grant proposals and reports, etc.

85/73. 1 carton. Additions.

Taub, Abraham H. (1911–   ), professor of mathematics, UCB.

82/136. 9 cartons, folder list. Subject files; correspondence; manuscripts of books, papers, and talks; course notes, etc. Includes correspondence with and about John von Neumann and materials concerning ENIAC.

Taylor, Walter P. (1888–1972), curator of the Museum of Vertebrate Zoology, UCB.

> 85/72. 5 cartons. Journals (ca. 1906–1967), including field notes.

Tchelistcheff, André (1901–   ).

> 84/172. Oral history, "Grapes, wine, and ecology."

Terman, Frederick (1900–1982), professor of electrical engineering, dean of the School of Engineering, and provost, Stanford University.

> Oral history (HSTP), including discussion of the Harvard Radio Research Laboratory, radar, Stanford University, electronics, and Silicon Valley.

Testi, Cesare.

> 3Ms QA33 T4. "Matematica miscellan...," including extracts from Boethius, Commandino, and Jordanus (17th-century Italy).

Thevet, André (1502–1590), cosmographer to the French king.

> Film Z-F2. Microfilm copy of "Le grande insulaire et pilotage d'André Thevet" (1586). Original in the Bibliothèque Nationale, Paris.

Thompson, Robert,

> Film 67/190. Microfilm copy of the record of a voyage to the West Indies (1699–1700), including navigation instructions. Original in the Admiralty Library, London.

Thomson, Sir John A. (1861–1933).

> 71/167. 1 portfolio (1898–1924). Letters to Thomson, manuscripts and corrected proofs, notes, etc., concerning biology.

Thornton, Robert L. (1908–   ), professor of physics, UCB.

> Oral history (HSTP), including discussion of accelerators, nuclear physics, Lawrence Berkeley Laboratory, and the Manhattan Engineer District.

Tibbetts, Frederick Horace (1882–1938).

C-B785. 1 box (1912–1938). Articles, reports, a scrapbook, and clippings relating to his work as a consulting engineer. Includes materials concerning water resources.

Tibbetts, Sydney A., industrial chemist and geologist.

67/78. 1 portfolio (1925–1938). Geological reports.

68/75. 2 boxes (1907–1957). Papers and reports pertaining to irrigation, wells, soil science, geology, and water rights.

Torrey, Harry Beal (1873–1970).

71/89. 4 boxes, 1 carton. Correspondence; lectures, bibliography, and other materials concerning history of science; research materials relating primarily to physiology; course materials; subject files; photographs and slides, etc. Also MsPRS T68 1935 2.23.

Townsend, Calvin Kirk

78/21. 1 carton (1924–1959). Correspondence and papers, including business records concerning Jennings Radio.

Treadwell, George A.

76/49. 1 box, 1 portfolio (ca. 1866–1910). Correspondence, including letters from John James Rivers concerning zoological and botanical specimens and letters from the American Association for the Advancement of Science.

Turner, Francis J. (1904–), professor of geology and geophysics, UCB.

84/70. 6 cartons, oversize materials, folder list. Laboratory notebooks, research data, and photographs; grant proposals and reports; manuscripts and reprints of writings, with related correspondence; subject files; department and course materials, etc.

Tuttle, Albert H.

73/122, no. 86. "The Harvard that I knew" (1916), including reminiscences of Asa Gray and Louis Agassiz.

Tyler, John G.
79/111. 4 boxes. Letters received, many concerning the Cooper Ornithological Club. Includes letters from Joseph Grinnell.

U.S. Work Projects Administration, California.
C-R19. 1 carton (1935–1939). Field notes, reports, maps, photographs, etc., for the anthropological survey of Orange Country. Includes materials concerning archæology and California Indians.

Ulloa, Don Antonio de.
Microfilm copy. Abstract of the journal of observations he and French astronomers made in Peru in 1736 (read to the Royal Society in 1746). Originals in the Admiralty Library, London.

Underhill, Robert McKenzie (1893–   ), secretary and contracting officer for the UC Regents.
79/26. Oral history concerning UC and Los Alamos laboratory; supplementary documents about contract negotiations. Restricted. See also Be.05 Los Alamos Scientific Laboratory.

83/154. 1 box. Correspondence and papers relating in part to Los Alamos.

Album of photographs of Los Alamos laboratory facilities.

Untermyer, Samuel.
74/106. 2 cartons, folder list. Subject files, photographs, some correspondence, notebooks, etc., concerning nuclear reactor technology.

Usinger, Robert L. (1912–1968), professor of entomology, UCB.
73/68. 29 boxes, 24 cartons, oversize drawings. Correspondence, course materials, manuscripts of writings and talks, reprints, field notes, scientific files, photographs, drawings, slides and film, personalia, clippings, note cards, maps, etc. Includes materials concerning Usinger's collection of books by Linnaeus.

84/48. Additions.

Van Dyke, Donald C. (1923–), research physician at Donner Laboratory, UCB.

Oral history (HSTP), including discussion with Hal O. Anger.

Varian Associates.

73/65. 6 cartons (ca. 1948–1972), folder list. Corporate records and reports, including financial statements, minutes of meetings of the board of directors, studies, and news releases concerning electronics research and production.

Vasseur, Albert.

78/176. Copy of unpublished typescript, "De la TSF à l'électronique: Histoire des techniques radioélectriques" (1974), covering the years 1842–1950. Includes discussion of radar, microwaves, and television.

Vaux, Henry J., professor of forestry, UCB.

85/45. 13 cartons. Correspondence and papers.

Villars, Donald S. (1900–), physical chemist.

74/192. 20 cartons (ca. 1930–1965), carton list. Correspondence, subject files, research notes, course materials, audiovisual materials, etc., some of it concerning the University of Minnesota and U.S. Rubber Company.

Voy, C. D.

C-E127. "Relics of the Stone Age found in California" (1874), with preface referring to Josah Whitney and Joseph LeConte.

Warren, Earl (1891–1974), chief justice, U.S. Supreme Court.

72/100, 73/156. Oral history series, "Earl Warren and Health Insurance" and "Earl Warren and the State Department of Public Health."

Weeks, Walter Scott (1882–1946), professor of mining, UCB.

C-B1040. 1 portfolio (1932–1946). Correspondence and

papers relating to articles on mining devices and to service as university marshal.

Weitbrecht, Robert H. (1920–), astronomer.

85/67. 5 cartons. Correspondence and papers concerning his work in astronomy and his invention of electronic devices to assist the hearing impaired.

Welch, Charles. C-R28. 1 box (ca. 1909–1912). Materials relating to Luther Burbank, including a letter from Burbank, a scrapbook about the spineless cactus, and promotional material.

Whinnery, John Roy (1916–), professor of electrical engineering and dean of the College of Engineering, UCB.

78/144. 10 boxes, 9 cartons, folder list. Manuscripts of papers, research notes, materials concerning professional organizations and university and department activities, and correspondence. Includes materials concerning vacuum tubes and microwaves.

White, Harvey E. (1902–), professor of physics, UCB.

81/169, no. 1. Autobiographical account of his career in spectroscopy and optics, including work for the Navy in World War II, directorship of Lawrence Hall of Science, and interest in radio.

Wickson, Edward James (1848–1923), professor of agriculture, UCB.

C-B1035. 1 box, 1 carton (1868–1923), finding aid. Correspondence, manuscripts and reprints of his writings, materials about Luther Burbank, clippings, reviews, and obituaries.

Williams, Arthur Robinson (1884–1961), associate professor of mathematics, UCB.

C-B992. 1 carton (ca. 1920–1961). Notes, reprints, and manuscripts of papers, including papers of B. C. Wong.

Williams, Robley C. (1908–), professor of biophysics and molecular biology, UCB.

73/7. 3 cartons, folder list. Correspondence, lecture notes, patents, subject files, etc., concerning electron microscopy, virology, and biophysics.

Winkler, Albert J. (1894–).

74/166. Oral history, "Viticultural research at the University of California, Davis, 1921–1971."

Wise, W. Howard, physicist.

75/50, no. 18. 60 p. (spring 1926). Notes of the lectures of Paul Epstein on the quantum theory of atomic systems.

Wood, William Maxwell (1809–1880).

77/100. Log (1844–1845) kept while serving as fleet surgeon of the Pacific Squadron. Includes information concerning physiology.

Woodworth, Charles William (1865–1940), professor of entomology, UCB.

67/124. 1 portfolio (1888–1939). Miscellaneous papers, including accounts, photographs, a manuscript of a speech, postcards, and publications.

Woodyard, John (1904–1981), professor of electrical engineering, UCB.

82/40, 82/133. 1 box, 22 cartons (ca. 1930–1971). Correspondence, research notes, patents, course materials, etc., concerning his dissertation work at Stanford University, association with Sperry Gyroscope, and teaching and research at the University of Washington and UCB. Includes materials relating to the klystron, the Radiation Laboratory, and Los Alamos.

84/18. Additions.

Oral history (HSTP), "Cyclotron and klystron development."

Wyckoff, Florence.

77/67, v. 3. Oral history (1975), including discussion of public health issues.

Yates, Lorenzo Gordin (1837–1909).

C-B472. 3 boxes, 1 carton, 1 v. (1852–1906). Correspondence, diaries (1871, 1882), subject files, notes, manuscripts of his writings, catalogs of his shell and fossil collections, and clippings. Includes materials concerning mineralogy and other aspects of natural history.

C-D525. 1 box (1887). Dictation, letter, and autobiographical sketch concerning his natural history collections, writings, work on the California Geological Survey, and membership in scientific societies.

Yerushalmy, Jacob (1904–1973), biostatistician.

74/199. 7 cartons, oversize materials (1948–1971), folder list. Correspondence, subject files, reports, research data, materials relating to the National Institutes of Health and the U.S. Public Health Service, manuscripts and reprints of writings, etc.

Yoakum, Franklin L. (1819-?), botanist and mineralogist, Confederate States of America.

P-O75. Dictation.

York, Herbert F. (1921–), professor of physics, UCB and UC San Diego; director of the University of California Radiation Laboratory, Livermore.

84/15. Oral history (HSTP), "Physics at Berkeley and Livermore, 1943–1952."

75/50, no. 19. Transcript of oral history interview, including discussion of Project Sherwood.

Eleven Nobel Prize winners, University of California, Berkeley (1969). From left: (standing) Edwin McMillan, Donald Glaser, Emilio Segrè, Wendell Stanley, Melvin Calvin; (seated) Charles Townes, Owen Chamberlain, Glenn Seaborg, John Northrop, Luis Alvarez, William Giauque. University Archives, The Bancroft Library.

Be.06

Water Resources Center Archives
410 O'Brien Hall
University of California
Berkeley, CA 94720
(415) 642-2666

The Water Resources Center Archives is an activity of the statewide Water Resources Center of the University of California, inaugurated in 1957. The scope of the collection is multidisciplinary and includes materials relating to engineering, social, legal, physical, biological, and economic aspects of water resources development and management. The period covered is, in general, 1890 to date, with a particular concentration on California and the West. A multivolume *Dictionary catalog* (Boston, 1970; supplement 1971–1978) provides author and subject entries for all materials.

Holdings include collections of personal papers and records of consulting engineers and engineering firms; publications and papers of governmental agencies; minutes, hearings, and reports of committees, boards, and commissions; reports and bulletins of utility companies, research institutes, and hydraulic laboratories; house journals and serial publications of associations and societies; brochures, campaign materials, speeches, and addresses; photographs, slides, maps, charts, drawings, and other illustrative materials; and newspaper clippings and news releases.

DAVIS

Da.01

Special Collections
Shields Library
University of California
Davis, CA 95616
(916) 752-1621

Major subjects of the archival and manuscript holdings at the University of California, Davis, include agricultural technology, history of agriculture, and history of the Davis campus, as well as dramatic art, contemporary American literature, and aspects of California history and culture. Consultation of materials in Special Collections generally requires at least three days' notice.

Agricultural technology.

012. 150 document cases (1885–1982), finding aid. Catalogs.

005. 3 cartons (1850–1959), carton list. Materials relating to agricultural technology, excluding manuals and catalogs.

011. 150 document cases (1885–), finding aid. Manuals for operation, maintenance, repair, or restoration of agricultural machinery.

American Beekeeping Federation.

D-74. 1 folio box, 11 cartons (1940–1971), carton list. Records, including correspondence, financial records, and printed materials.

Amerine, Maynard Andrew (1911–), professor of viticulture and œnology, UCD.

D-60, D-21. 1 folio box, 12 cartons, 2 document cases (ca. 1925–1978), unprocessed. Correspondence and papers, including 10,000 wine bottle labels. See also D-128.

Apiculturalists.

D-76. 1 carton (ca. 1880–1958), unprocessed. Manuscripts, photographic images, and printed materials relating to 19th-century apiculturalists and apiaries.

Apiculture subject files.

D-77. 1 folio box, 3 cartons (ca. 1860–1975), unprocessed. Manuscripts, photographic images, and printed materials relating to pollination, hives, diseases, pesticides, etc.

Audio-visual.

014. 10 cartons (1931–1982), finding aid. Cassettes, films, tapes, and videotapes relating especially to UCD faculty, agriculture, apiculture, and wine.

Bailey, Stanley F., professor of entomology, UCD.

D-98. 4 cartons (ca. 1945–1960), unprocessed. Correspondence and papers.

Bayard, Arnold A.

D-05. 2 cartons (1869–1981), unprocessed. Wine catalogs and price lists for French wines.

Beekeeping supply catalogs.

D-78. 22 pamphlet boxes, 1 carton (1880–1978), finding aid.

Bohart, Richard, professor of entomology, UCD.

D-99. 1 carton (ca. 1945–1960), unprocessed. Correspondence and papers.

Born, Leonard L.

D-85. 5 cartons (1949–1968), carton list. Research files on the radiation and cold sterilization of foods.

Botanical prints.

D-42. 4 folio boxes (ca. 1960), unprocessed. Prints by Henry Evans and Edmund E. Simpson.

Brooks, Frederick A., professor of agricultural engineering, UCD.

D-100. 4 cartons (ca. 1945–1960), unprocessed. Correspondence and papers.

Butterfield, Harry Morton (1887–), agriculturalist, UCB.

D-86. 4 cartons (1860–1971), unprocessed. Papers on the history and introduction of ornamental plants and plantings in California.

California State Beekeepers Association.

D-75. 6 cartons (1891–1973), inventory. Organizational materials, correspondence, minutes, and printed materials.

California wineries.

D-140. 171 cartons (1920–1947), carton list, unprocessed. Records relating to the California wine industry as held by the U.S. Bureau of Alcohol, Tobacco, and Firearms, covering the period of Prohibition and its repeal.

Caplan, Herb.

D-73. 2 cartons (1889–1979), unprocessed. 416 pamphlets on varnish.

Čebiš, František Rudolf.

D-123. 4 folio boxes, 15 cartons (1801–1960), unprocessed. Chiefly wine bottle labels from eastern Europe; advertising brochures, pamphlets, and clippings relating to wine.

Corti, Egon Caesar, conte (1886–1953).

D-10. 3 folio boxes, 2 cartons (1805–1912), inventory. Lithographs, drawings, books, and manuscript notes used for 19th-century German pomological texts.

Cove/Bakken Advertising Company.

D-4. 50+ cartons (ca. 1955–1982), unprocessed. Archives of an advertising company promoting agricultural machinery.

Crafts, Alden Springer (1897–), professor of botany, UCD.

D-87. 13 cartons (1915–1965), carton list. Papers relating to chemical weed control research employing autoradiography.

Craig, Thornton, physician.

D-88. 50 volumes (1876–1922), inventory. Holograph diaries, relating mainly to visits, prescriptions, and accounts.

Cruess, William Vere (1896–1968), professor of food science and technology, UCB.

D-53. 2 cartons (1923–1968), inventory. Manuscripts, correspondence, and photographic images relating to work on canning, preserving, wine, and winemaking.

Decker, Frank Norton, attorney.

D-90. 1 document case (1965–1966). Carbon copies of manuscripts, "To: National Advisory Commission on Food and Fiber" and "In re natural law." Includes references to Cornell University.

Eckert, John Edward (1895–), professor of entomology, UCD.

D-79. 18 cartons (1936–1958), inventory. Chiefly correspondence between Eckert and international apiculturalists (1951–1958).

Edison, Thomas Alva (1847–1931), inventor.

D-52. Holograph letter (1894) regarding mining and Major McLaughlin, written on carbon typescript (1883).

Esau, Katherine (1898–), professor of botany.

D-120. 1 carton (1930–1965). Reprints of her publications.

Faculty reprints.

D-125. 163 cartons (1920–1980), arranged by author. Includes reprints of publications by UCD faculty members concerning agriculture, botany, chemistry, genetics, and earth sciences.

Fairbanks, James P., agriculturalist, UCB.

D-101. 7 cartons (ca. 1945–1960), unprocessed. Correspondence and papers.

Federal wine bottle labels.

D-127. 26 cartons (1963–1968), arranged alphabetically. Labels as submitted to the Internal Revenue Service by liquor dealers and importers.

Ferry-Morse Seed Company.

D-114. 4 folio boxes, 29 cartons (1870–1954), inventory. Seed catalogs, original art work, and field notes.

Freeborn, Stanley Barron (1891–1960), provost, UCD; entomologist.

D-102. 1 carton (ca. 1946–1955), unprocessed. Correspondence and papers, including materials pertaining to medical entomology and to mosquitos.

Gilmore, Arthur E.

D-92. 2 cartons (1955), unprocessed. Photographs, etc., relating to flood control and geology.

Guymon, James Fuqua (1911–1978), professor of œnology, UCD.

D-115. 27 cartons (1938–1978), folder list. Papers concerning wine and brandy technology, distillation, refrigeration, biochemistry of alcoholic fermentation, and aging phenomena.

Harbison, John Stewart (1826–1912), beekeeper.

D-80. 1 carton (1857–1912), inventory. Papers of one of the first beekeepers to import bees into California.

Hart, George H. (1888–1959), dean of the School of Veterinary Medicine, UCD.

D-103. 1 carton (ca. 1950–1955), unprocessed. Correspondence and papers.

Higgins, Floyd Halleck (1886–1975), journalist; news editor for Caterpillar Tractor Company.

D-56. 427 cartons (1850–1959), finding aids, partially unprocessed. Correspondence, manuscripts, brochures and circulars, patents and contracts, bills and receipts, house publications, graphic material, clippings, ephemera, artifacts, and photographs concerning the farm industry, especially agricultural technology and machinery.

D-116. 23 cartons (1927–1975), unprocessed. Papers concerning his research on primary agricultural industries and their history, including agricultural technology.

Holland Land Company.

D-118. 16 ledgers (1893–1945), unprocessed. Minutes of the company in Reclamation District 551, Northern California.

Hunt, Thomas Forsyth (1862–1927), dean of the College of Agriculture, UCD.

D-104. 1 carton (ca. 1910–1920), unprocessed. Correspondence and papers.

Hutchison, Claude Burton (1885–1980), university vice president, UC.

D-105. 3 cartons (ca. 1924–1927), unprocessed. Chiefly log books for agricultural education projects in Europe.

Index seminum.

D-25. 1 carton (1965–   ), finding aid. Seed lists from foreign and American arboreta and botanical gardens.

Jacobsen, Henry, agricultural inspector in the Philippines.

D-21. 1 carton (1912–1916), unprocessed. Lantern slides and scrapbooks.

Kleiber, Max (1893–1976), professor of animal husbandry, UCD.

D-107, D-30. 35 cartons (ca. 1925–1977), carton list. Papers, primarily concerning metabolic research in animal physiology.

Laidlaw, Harry Hyde, Jr. (1907–), professor of entomology, UCD.

D-81. 1 carton (1948–1951), inventory. Correspondence, some concerning artificial insemination of queen bees.

Liberty Farms Company.

D-44. 4 folio boxes, 1 carton (1916–1974), finding aid. Archives of California Delta region farming company owned by Robert K. Malcolm (1868–1951), a pioneer in large-scale farming.

Madson, Ben A., professor of agronomy and director of field stations, UCD.

D-108. 1 carton (ca. 1945–1960), unprocessed. Papers.

Marsh, Warner, and Florence Marsh.

D-72. 2 cartons (ca. 1919–1960), unprocessed. Collection of bulletins, chiefly concerning forestry.

Maillard family.

D-18. 3 cartons (1941–1973), unprocessed. Records of breeding experiments with Merino sheep imported from New Zealand, including ewe production, flock, wool, and lamb records.

McCubbin, John Cameron (1863–?), California beekeeper.

D-82. 2 cartons (1891–1928), carton list. Chiefly correspondence.

Miller-Lux.

D-94. 14 cartons (ca. 1860–1890), carton list. Archives of the early farming partnership of Henry Miller and Charles Lux in California.

Morrow, Dwight W., Jr. professor of history, Monterey Institute of Foreign Studies.

>D-95. 21 card files, 2 cartons (1861–1954), carton list. Papers, chiefly concerning *phylloxera* disease in grapes and Franco-American agricultural history, especially the history of viticulture in France.

Mrak, Emil Marcel (1901–), food scientist and chancellor, UCD.

>D-96. 49 cartons (1940–1978), restricted, folder list. Papers, many concerning pesticides, herbicides, and world agricultural development.

Nelson, Oscar M. F.

>D-119. 1 carton (ca. 1925–1931), unprocessed. Manuscripts and mimeographed papers concerning sugar and methods of refining sugar.

Neubauer, Loren W., professor of agricultural engineering, UCD.

>D-109. 1 carton (ca. 1950–1960), unprocessed. Papers.

Nursery and seed catalogs.

>009. 230 linear feet (1850–), finding aid.

Peterson, Maurice L., professor and chairman of agronomy, UCD.

>D-118. 1 carton (1950), unprocessed. Engravings for a publication concerning Sudangrass.

Photographs.

>007. 50,000 items (1850–1982), unprocessed. Great variety of photographic images, including those depicting agricultural machinery and most phases of the agricultural industry.

Pomology glass slides.

>D-126. 1 cabinet (1905–1951), finding aid. Slides concerning pomology and fruit processing.

Portuguese wine pamphlets.

D-122. 5 cartons (1801–1952), finding aid. Pamphlets (mostly in Portuguese) relating to the growing, manufacture, and economics of grapes and wine.

Power/Drake research.

D-19. 3 cartons, oversize materials (1962–1979), inventory. Correspondence and papers relating to the 400th anniversary of Sir Francis Drake's voyage around the world and the time he spent in California in 1579.

Richardson, Sir Benjamin Ward.

D-129. 1 carton (1745–1902), unprocessed. Papers, diagrams, and correspondence relating to the Richardson Lethal Chamber, an early gas death chamber used in animal research.

Richter, Max Clemens (1884–), botanist and commercial beekeeper.

D-83. 1 carton (1887–1954), inventory. Correspondence and papers.

Rowe, Albert H., lecturer in medicine, UCB.

D-14. 11 cartons, unprocessed. Correspondence and papers, including materials concerning research on allergies.

Ryerson, Knowles Augustus (1892–), dean of the College of Agriculture, UCD.

D-11. 27 cartons (1892–1965), folder list. Correspondence and papers, including materials concerning agricultural education.

Scheuring, Ann Foley.

D-48. 1 document case (ca. 1960–1980), unprocessed. Photographs relating to California agriculture, some of which were used in a *Guidebook to California agriculture.*

Shields, Peter J. (1862–1962), Superior Court Judge, California.

D-130. 3 cartons (1908–1961), finding aid. Papers,

correspondence, and memorabilia of Judge Shields, who founded University Farm, now the University of California, Davis.

Skaar Mining Company.
D-16. 2 folio boxes, 4 cartons (1786–1959), carton list. Archives relating to the development of mining in the California gold fields.

Stebbins, George Ledyard (1906–), professor of genetics, UCD.
D-117. 1 carton (1970–1982). Correspondence files, including materials concerning botany and cytogenetics.

Steinmetz, Andrew.
D-97. 1 box. Two manuscripts, "The weather and our food-prospects" and "The duration of life and average amount of sickness of the various trades."

Storer, Tracy Irwin (1889–1973), professor of zoology, UCD.
D-110. 10 cartons (ca. 1940–1965), unprocessed. Correspondence and papers, including materials concerning public health.

University Archives.
AR 1– . 300+ cartons (1906–), unprocessed. Academic and administrative papers concerning all departments and units within the University of California, Davis.

Vermouth project.
D-128. 3 cartons (ca. 1970), unprocessed. Papers concerning Dr. Amerine's vermouth project. See also D-21, D-60.

Voorhies, Edwin Coblentz (1892–1967), professor of agricultural economics, UCD.
D-89. 4 cartons (1945–1962), carton list. Papers relating to agricultural economics, the Department of Agricultural Economics at Davis, and translations of *Aereboe Agrarpolitik*.

Walker, Harry Bruce (1884–1957), professor of agricultural engineering and department chairman, UCD.

D-111. 7 cartons (ca. 1930–1955), unprocessed. Papers, including materials relating to his role in the development of the sugar beet harvester.

Watkins, Lee H. (1908?–1972), agricultural assistant, UCD.

D-84. 8 cartons (1939–1972), unprocessed. Correspondence relating to apiculture and research papers about the history of beekeeping in America.

Wickson, Edward James (1848–1923), dean of the College of Agriculture, UCD.

D-37. 1 carton (1874–1919), unprocessed. Papers and correspondence of Wickson with or about Luther Burbank.

Wine newsletters.

D-124. 5 cartons (1939–1976), unprocessed. Collection of winery and vineyard newsletters, with an emphasis on California.

Wines and vines.

D-71. 4 cartons, oversize materials (1970–1979), unprocessed. Primarily photographs used in *Wines and vines*, plus wine labels.

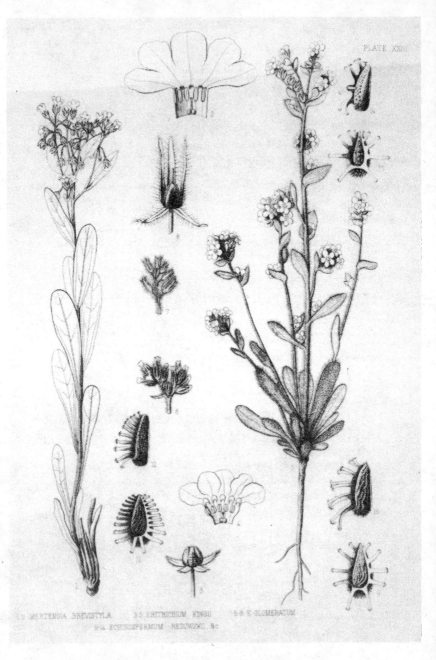

Botanical plate (1871), published in connection with the U.S. Geological Exploration of the 40th Parallel directed by Clarence King. The Bancroft Library.

SACRAMENTO

Sa.01

California State Archives
1020 O Street, Room 130
Sacramento, CA 95814
(916) 445-4293

The holdings of the California State Archives, a division of the Office of the Secretary of State, amount to 55,000 cubic feet of state government and country records, photographs, maps, microforms, manuscripts, and printed books. Among the records pertaining to science and technology are reports of the State Geological Survey and the State Geologist and Mineralogist for the 19th century and records of such departments and agencies as the Departments of Health and Agriculture, the Office of Appropriate Technology, and the Air Resources Board for the 20th century. The records reflect the variety of scientific and technological issues addressed by state agencies. The papers of the Department of Health, for example, concern food technology, epidemiology, sanitary engineering, water resources, and other matters affecting public health; those of the Department of Agriculture deal with disease control, entomology, and ecology as well as botany and agricultural technology. Other archival collections of interest include census records, court and legislative files, and some gubernatorial papers. Inventories are available for some record groups; access to certain records is closed or restricted.

Sa.02

California State Library, California Section
Library-Courts Building
P.O. Box 2037
Sacramento, CA 95809
(916) 445-2585

Among the holdings of the California Section of the State Library are many manuscript collections concerning railroads and mining, some of which are described below. Readers are urged also to consult the California Section information file, which indexes selected materials from books, periodicals, and newspapers, as well as manuscript collections in the Library. Unprocessed collections are restricted and may be used only with special permission.

Carson Hill Gold Mining Corporation.

Box (carton) numbers 858–877 (1917–1942), collection summary. Corporation correspondence and records.

Givan, Albert.

9 cartons, charts, restricted. Maps, water reports, engineering studies, etc., many relating to the Sacramento Municipal Utility District.

Hall, William Hammond (1846–1934), hydrographer; California state engineer.

1133–1142, collection summary. Correspondence, business and personal papers, notebooks, scrapbooks, etc. Includes materials concerning mining in California, Africa, and Russia; irrigation; and water resources.

Hendel, Charles William (1831–1920), surveyor and miner.

661–701 (1853–1918), collection summary. Business, legal, and mining papers, including field notes and surveys.

Hiller, Stanley, Sr. (1888–1968), inventor and aviator.

> 501–644 (1922–1962), collection summary. Business and personal papers, including materials concerning food processing technology. Plus some papers of Stanley Hiller, Jr., concerning his new die-casting process and his invention of the first practical coaxial helicopter.

McCartney, Henry Maxwell (1848–1915), railroad and survey engineer.

> 834–853 (1869–1911), collection summary. Business papers, charts, correspondence, diaries, maps, photographs, and railroad papers. Includes materials concerning mining and surveying.

McIntire, John Albert (1843–1931).

> 364, 393–408, 498 (1880–1931), collection summary. Business papers, clippings, correspondence, mining information, notes and drafts, photographs, and printed materials, many concerning his consolidation of mining companies along the Mother Lode.

Mining companies miscellany.

> 930–933 (ca. 1900–1930), collection summary. Reports on mining companies, including maps, photographs, and broadsides; pamphlets, attorneys' briefs for court cases, and surveyors' notebooks.

Morse, Ephraim W. (1823–1906), businessman and county official.

> 24874, 382–392 (ca. 1849–1905), collection summary. Account books, business and legal papers, correspondence, Spanish and Mexican documents relating to mines in Baja California; express and shipping receipts.

Moss, Joseph Mora (1809–1880), California businessman.

> 1148–1155 (1849–1915), collection summary. Business and personal papers, such as receipt, record and ledger books; photographs; plus papers of Dr. Joseph Mora Moss III. Includes materials concerning a predecessor of Pacific Gas and Electric Company, the California State Telegraph Company, and the Market Street Railroad.

National Bank of D. O. Mills & Company.

368–378 (1849–1927), collection summary. Letters, deeds, stocks and bond, pictures, estate papers, checks, and other bank items, some concerning mining and railroads.

Natomas Company and its predecessors.

cTN423 Z6 N25. 29 v. (1851–1956). Articles of incorporation and other business and legal records, mining and crop statistics, clippings, maps, pamphlets, photographs, etc. Includes materials relating to mining, land reclamation, agriculture, and water resources.

New Almaden Mine.

6 cartons, 1 box, charts (ca. 1925–), restricted. Business papers, correspondence, etc.

New Idria Mining Company.

275 (1854–1867), collection summary. Business records and correspondence.

Sutro, Adolph (1830–1898), mining engineer; mayor of San Francisco.

Drawers 1–42, collection summary. Correspondence, deeds, ephemeral materials, pamphlets, printed materials, receipts, typescripts, vouchers, etc. Includes materials relating to railroad funding, book collecting, and the Sutro Tunnel project to drain mines in the Comstock Lode.

U.S. Land Office, Sacramento.

815–822 (1858–1908), collection summary. Correspondence, deeds, legal records, and mining papers.

Winchester, Jonas (1810–1887), businessman and journalist.

329–338 (1829–1917), collection summary. Correspondence, reflections, books, pictures, copies of newspapers, etc. Includes materials relating to mining.

SAN FRANCISCO

SF.01

Archives
California Academy of Sciences
Golden Gate Park
San Francisco, CA 94118
(415) 221-5100, ext. 260

The Archives' holdings deal primarily with the history of the California Academy of Sciences, founded in San Francisco in 1853. The emphasis is on the post-1906 period; most earlier correspondence and archival materials perished in the earthquake and fire in San Francisco in 1906. Other holdings include correspondence and papers, dating from the 19th and 20th centuries, of such naturalists and scientists as J. G. Cooper, George Davidson, Alice Eastwood, Gustav Eisen, and B. W. Evermann. Access to the Archives is by appointment only.

SF.02

Department of Special Collections
The Library
University of California
San Francisco, CA 94143
(415) 666-2334

In addition to historical and administrative records of the medical facilities and schools that constitute the University of California at San Francisco, the holdings of the Department of Special Collections include rare books and manuscript collections relating to the health sciences.

Arnot, Philip H. (1894–1974), chief of obstetrics and gynecology, St. Mary's Hospital.

> MSS 74-11. 5 cartons, 1 oversize box (1908–1974), folder list, patient records restricted. Correspondence, research materials (including illustration and demonstration materials), manuscript notes on medical school lectures, etc.

Audy, J. Ralph (1914–1974), director of the Hooper Foundation.

> MSS 75-9. 11 cartons (1959–1974), carton summaries. Correspondence, course syllabi, notebooks, lectures, grant applications, etc., concerning such topics as epidemiology, ecology, preventive medicine, and the British Colonial Office Scrub Typhus Research Unit, Kuala Lumpur.

Berne, Eric (1910–1970), psychoanalyst; founder of transactional analysis.

> MSS 82-0. 2 boxes, 3 cartons (1959–1966), finding aid. Audio and videotapes, photographs, reprints, a heavily annotated 350-volume working library, etc. Includes materials concerning the Transactional Analysis Institute of San Francisco.

Brigham, Charles Brooks (1845–1903), professor of surgery, UCSF.

2 boxes (1870–1902), folder list. Notes, manuscripts, correspondence, case histories, galleys, photographs, etc., concerning such topics as surgery, gastrectomy, the Franco-Prussian war, orchids, and the Ambulance de l'Ecole Forestière.

Burbridge, Thomas N. (1921–1972), associate professor of pharmacology and experimental therapeutics, UCSF.

4 cartons (1959–1972), folder list. Lecture notes, correspondence, manuscripts, photographs, etc., relating to research in comparative pharmacology, alcohol metabolism, and psychopharmacology.

California Society for the Promotion of Medical Research.

1 box, 3 cartons (1936–1938). Papers concerned with finances, advertising, and publicity relating to the passage of the humane pound act, etc.

Cullen, Stuart Chester (1909–1979), professor of anesthesiology, UCSF.

2 cartons (1931–1960), folder list. Typescripts, correspondence relating to publications, lecture notes, research notes (1942–1949) on anesthesia responses and curare, etc.

Dental oral history project.

1 carton (1967–1974). Tapes, transcripts, biographical materials, and photographs for 16 dentists from the San Francisco Bay Area.

Dixon, Robert E. (1861–1932), physician in Hanford, California.

MSS 84-5. 2 boxes (1887–1932), folder list. Correspondence, account books, newspaper clippings, photographs, phrenological charts, and certificates relating to medical practice in the San Joaquin valley.

Dunphy, John E. (1908–1981), professor of surgery, UCSF.

MSS 82-8. 9 cartons (1974–1981), folder list. Correspondence, photographs, manuscripts, notes and typescripts of lectures, papers, referee reports, etc., some relating to the Veterans Administration Hospital in San Francisco and to cancer surgery.

Fleming, Willard C. (1899–1973), dean of the School of Dentistry and chancellor, UCSF.

MSS 73. 5 cartons (ca. 1940–1969), carton list. Correspondence, appointment books, photographs, etc.

Fraiberg, Selma H. (1918–1981), professor of child psychoanalysis, University of Michigan and UCSF.

MSS 83-9. 22 cartons (1940–1981), carton summaries; some restrictions may apply. Correspondence, teaching files, manuscripts and typescripts, drafts and galleys of books, grant applications, audio tapes, photographs, etc. Includes materials relating to psychoanalytic institutes in Baltimore and Michigan.

Harris, Henry (1874–1938), physician, lecturer in medical history, UCSF.

6 boxes (1926–1938), folder list. Typescripts; manuscripts and notes for *California's medical story* and for an unfinished history of bacteriology in California, begun with K. F. Meyer; a portfolio of original musical compositions, etc. Includes materials relating to Stanford.

George Williams Hooper Foundation.

AR 59-1. 45 cartons (1882–1958), card index. Records, including correspondence, financial records, and pamphlets, concerning such subjects as the California canning industry, leptospirosis, botulism, immunity, animal husbandry, antivivisection, epidemiology, zoonoses, and tropical medicine.

Inman, Verne (1905–1980), professor of orthopedic surgery, UCSF.

MSS 80-6. 9 boxes, 3 cartons (1950–1980), folder list; patient records restricted. Correspondence, slides, photographs, models, and notes relating to lectures, meetings, the Biomechanics Laboratory, and orthopedic research on movement and muscle function, etc.

Junck, Anita C., dental hygienist and UCSF graduate.

> MSS 83-12. 1 oversize box (ca. 1923–1940), folder list. Materials concerning dental hygiene for children, including a scrapbook and loose forms, advertisements, illustrations, pamphlets, handbills, and official notices from departments of public health and hygiene in the United States and Canada.

Olmsted, Evangeline Harris.

> 2 boxes (1928–1975), folder list. Correspondence, manuscripts, memorabilia, reprints, typescripts, diaries, notebooks, etc., relating to J. M. D. Olmsted, Claude Bernard, and her own activities.

Olmsted, James M. D. (1886–1956), professor of physiology, UCSF and UCB.

> MSS 78-1. 3 cartons (1896–1956), folder list. Correspondence, lectures, manuscripts, memorabilia, notebooks, etc., relating to physiological research and historical research on Claude Bernard, François Magendie, and others.

Pacific Coast Surgical Association.

> 2 boxes, 4 cartons (1926–), folder list. Records, including correspondence, financial records, annual meeting programs, membership records, photographs, and slides.

Patt, Harvey (1908–1982), professor of radiobiology and director of the Radiobiology Laboratory, UCSF.

> MSS 83-5. 5 cartons (1949–1982), folder list. Business and personal correspondence, committee correspondence, minutes of meetings, reports, photographs, slides, appointment books, etc. Includes materials concerning Argonne National Laboratory and research in cytokinetics, radiobiology, and experimental hematology.

Riegelman, Sidney (1921–1981), professor of pharmacy and pharmaceutical chemistry, UCSF.

> MSS 82-12. 13 cartons (1960–1981), summary folder list. Correspondence, manuscripts, lecture notes, laboratory notes, records, computer printouts, etc., relating to research in pharmaceutical chemistry and pharmacokinetics, faculty matters, and teaching.

Rosencrantz, Esther (1876–1950), associate professor of medicine, UCSF.

MSS 51-1. 2 cartons (ca. 1920–1950), folder list. Manuscripts and typescripts, pamphlets, reprints, photographs, etc., relating to tuberculosis, the Golden Gate International Exposition, faculty matters, Johns Hopkins, and her interest in Osler and Garrison.

Sigma Theta Tau, Alpha Eta Chapter.

1 carton, 1 box (1964–), folder list. Chapter records, including correspondence, financial reports, program and publicity reports, membership information, and convention accounts, all relating to nursing.

Sovary, Lily.

MSS 83-12. 2 boxes (ca. 1965–). Personal accounts, artwork, correspondence, etc., some concerning Cushing's and Addison's diseases.

Strait, Louis A. (1907–1975), professor of biophysics and pharmaceutical chemistry, UCSF.

6 cartons (1943–1970), folder list. Correspondence, committee minutes, etc., relating to teaching, the Spectrographic Laboratory, and the Academic Senate.

West Bay Health Systems Agency.

95 cartons (ca. 1976–1982), unprocessed. Correspondence, chronological, data, subject, and agency files, etc.

SANTA CRUZ

SC.01

Mary Lea Shane Archives of the Lick Observatory
McHenry Library, Room 359
University of California
Santa Cruz, CA 95062
(408) 429-2571

Lick Observatory, constructed on Mount Hamilton according to provisions in James Lick's will, became part of the University of California in 1888. Its archives, now housed in the UCSC Library, bear the name of Mary Lea Shane in recognition of her long association with the Observatory and her dedication to the task of preserving and arranging the documentary record of the institution and its staff. In addition to the collections described below, the Archives contain correspondence and papers of many other astronomers, including E. E. Barnard, S. W. Burnham, Henry Crew, Heber D. Curtis, W. J. Hussey, G. W. Richey, and Joel Stebbins.

Aitken, Robert Grant (1864–1951), director of Lick Observatory (1930–1935).

1 linear foot (1893–1950), American Institute of Physics (AIP) finding aid. Scientific and personal correspondence, Lick Observatory records, reports, and papers.

Campbell, William Wallace (1862–1938), director of Lick Observatory (1901–1930) and university president, UC.

9 boxes plus 3 linear feet (ca. 1885–1938), AIP finding aid. Scientific and personal correspondence; correspondence and papers concerning Lick Observatory, the University of California, and eclipse expeditions; reports and papers.

Floyd, Richard S. (1843–1890), president of the James Lick Trust.

> 1 box (ca. 1875–1888), AIP finding aid. Correspondence with Thomas E. Fraser, Edward S. Holden, trustees of the James Lick Trust, scientific advisors, and manufacturers of telescopic equipment.

Fraser, Thomas E. (1842–1891), superintendent of construction for Lick Observatory.

> 1 box (1876–1889), AIP finding aid. Correspondence with Richard S. Floyd and Edward S. Holden, reports, and diary of the construction of Lick Observatory.

General collection.

> Correspondence, diaries, photographs, portraits, newspaper clippings, memorabilia, etc., concerning the construction, staff, equipment, buildings, and history of Lick Observatory. Includes materials concerning other observatories and astronomical meetings, together with personal collections of some individual astronomers.

Holden, Edward Singleton (1846–1914), director of Lick Observatory (1888–1897) and university president, UC.

> 3 boxes plus 13 linear feet (1874–1911), AIP finding aid. Correspondence with Richard S. Floyd, Thomas E. Fraser, Lick trustees, other scientists, manufacturers of telescopic equipment, etc.; records and papers concerning Lick Observatory and the University of California; reports; papers; technical drawings; and personal correspondence. Includes 87 wet-process copybooks.

Keeler, James Edward (1857–1900), director of Lick Observatory (1898–1900).

> 1 box (1875–1900), AIP finding aid. Correspondence with Edward S. Holden, W. W. Campbell, and other astronomers; reports; papers; and drawings of planets.

Leuschner, Armin Otto (1868–1953), director of the Students' Observatory and dean of the Graduate School, UCB.

> 3 linear feet (1889–1944), AIP finding aid. Correspondence with W. W. Campbell and other astronomers and faculty.

Equatorial telescope at Lick Observatory (1888). University Archives, The Bancroft Library.

Moore, Joseph Haines (1878–1949), director of Lick Observatory (1942–1946).

1 linear foot (1906–1947), AIP finding aid. Correspondence with other astronomers, Lick Observatory records, papers, and reports.

Photographic collection.

25 linear feet (ca. 1880–present), AIP finding aid. Photographs of the construction, staff, telescopes, and buildings of Lick Observatory, and of other observatories, astronomers, equipment, and professional meetings. Includes a unique 11-volume collection of cabinet portraits and engravings of early astronomers and other men and women of science.

Shane, Charles Donald (1895–1983), director of Lick Observatory (1945–1958).

14 linear feet (1916–1982), AIP finding aid. Correspondence, autobiography, papers and reports, descriptions of site selection trips, oral history interviews, photographs, and memorabilia.

Wright, William Hammond (1871–1959), director of Lick Observatory (1935–1942).

1 linear foot (1905–1955), AIP finding aid. Correspondence with other astronomers, Lick Observatory records, papers, and reports.

SC.03

McHenry Library
University of California
Santa Cruz, CA 95064
(408) 429-2970

Among the special collections at the UCSC Library are the South Pacific Collection of some 8500 titles and related manuscript materials in the South Pacific Archives. At the heart of the Archives are the papers of the South Pacific Commission, a consultative and advisory body established in 1947 by the six governments then responsible for the administration of island territories in the South Pacific region. The purpose of the Commission is to advise the participating governments regarding health, economic, and social conditions in the island territories. The records and publications of the Commission, as preserved in the South Pacific Archives, reflect the Commission's concern with many issues involving science, medicine, and technology.

The Archives also contain records, gathered by Howard G. McMillan and Knowles A. Ryerson, of the U.S. Commercial Company. During World War II the Company bought strategic materials from neutral countries and developed island agricultural resources in order to aid the war effort. McMillan's files, which span the years 1943–1952, contain records of the research activities of the Company, including its maps of Micronesia and an economic survey of the Central Pacific Islands in 1946–1947. Ryerson's papers include records of the Pacific Science Board of the National Academy of Sciences, which he chaired from 1946 to 1954. The Board was intended to aid American scientists in investigations in the Pacific region, to advise agencies on scientific issues pertaining to the Pacific, and to promote international cooperation in Pacific science. One of the projects of the Pacific Science Board was the Coordinated Investigations of Micronesian Anthropology (CIMA) in 1947–1949. Papers concerned with CIMA, from McMillan's files, are also preserved in the South Pacific Archives. Official records of CIMA for the years 1949–1953 are available on microfilm in the UCSC Library, as are the records of the U.S. Commercial Company. Finding aids have been prepared for both the South Pacific Archives and the South Pacific Collection of published materials.

STANFORD

The rare book, manuscript, and circulating collections of the Stanford University Libraries jointly support Stanford's rapidly growing program of research and instruction in history of science and technology. These collections are housed primarily in the Cecil H. Green Library, with additional historical collections in the rare book collection of the Branner Earth Sciences Library, Lane Medical Library's History of Medicine Collection, and the library of the Stanford Linear Accelerator Center.

Collections at Stanford contain rare books and supporting materials pertaining to the development of science and medicine from roughly 1500 to 1800, and materials concerning the growth of science and science-based technology since 1800, especially in the 20th century. Many of the rare book holdings are located in the Samuel I. and Cecile M. Barchas Collection in the History of Science and Ideas, the Frederick E. Brasch Collection on Sir Isaac Newton and the History of Scientific Thought, the Stephen P. Timoshenko Collection in the History of Mechanics, the former collection of the William Russell Dudley Herbarium, and the History of Medicine Collection in the Lane Medical Library. Collections of research journals, a unique collection of technical reports in computer science, engineering, and physics, and the high-energy physics collection of the Stanford Linear Accelerator Center supplement extensive book and manuscript holdings in disciplines ranging from fundamental physics to applied earth sciences, from artificial intelligence to biotechnology.

As part of a national project to preserve library resources published on now brittle paper between 1870 and 1920, Stanford is responsible for microfilming materials in the physical sciences published in the United States. By 1987 a total of nearly 6000 volumes, including runs of scientific journals not available from commercial distributors, will have been filmed and archived under this project. Stanford also plays a leading role in the creation of a national database of archival collections, including many of interest to historians of science and technology, supported by the Research Libraries Information Network. Stanford has also undertaken a survey of available resources for historical and sociological study of research and development in science-based technology, both at Stanford and in the industrial firms of Santa Clara Valley. For further information, write or call the Bibliographer for History of Science and Technology Collections, Stanford University Libraries, (415) 497-4342.

St.01

University Archives
Cecil Green Library
Stanford University
Stanford, CA 94305
(415) 497-2952

Abrams, Leroy (1874–1956), professor of biology and botany, Stanford.

SC13. 1 box (1863–1939), folder list. Correspondence, scrapbook, photographs, and personal papers.

ACME Project.

SC236. 6 linear feet (1961–1973). Files relating to the Advanced Computer Medical Experiments (ACME) Project at Stanford University Medical Center, to the use of computers, and to the establishment of the Stanford Computing Center. Compiled by Joshua Lederberg.

Alway, Robert, dean of the Medical School and medical director of Stanford University Hospital.

Tape 4:1. Oral history.

Bennett, Merrill Kelly (1898–1969), director of the Food Research Institute, Stanford.

SC149. 2.5 linear feet (1917–1966), folder list. Primarily professional and personal correspondence; unpublished writings (chiefly speeches) on food research and related topics; and correspondence regarding travel and research in Africa, Hawaii, and Japan.

Berg, Paul (1926–), professor of biochemistry, Stanford; Nobel laureate.

RGPN RG36. Videocassette recording of press conference upon receipt of the Nobel Prize (October 1980).

Berry, Samuel Stillman (1887–1984), zoologist.

Oral history, including discussion of marine biology.

Bloch, Felix (1905–1983), professor of physics, Stanford; Nobel laureate.

35 boxes (ca. 1934–1983), restricted, finding aid in progress. Research notes, notebooks, unpublished writings, reprints, some experimental data, and photographs of apparatus; professional correspondence, pertaining especially to his directorship at CERN and presidency of the American Physical Society.

Botanical Society of Leland Stanford Junior University.

SC158:6. Secretary's minutes and some treasurer's reports (1907–1908).

Bourguin, Burnice.

SCM002. Reminiscences regarding Frederick E. Terman as chairman of the Department of Electrical Engineering.

Bracewell, Ronald Newbold (1921–   ), professor of electrical engineering, Stanford.

1 v. Photocopy of manuscript, "Trees on the Stanford campus."

Branner, George Casper (1890–1967?), Arkansas state geologist.

SC35. 2 boxes (1896–1920). Papers relating to his childhood at Stanford as the son of its president, his experiences as a Stanford student in electrical engineering, and his experiences in World War I.

Branner, John Casper (1850–1922), professor of geology and university president, Stanford.

SC34. 93 boxes (1882–1921), folder list. Correspondence, manuscripts, field notebooks, photographs and line drawings of geological formations, slides, scrapbooks of earthquake photographs, reports, and published works. Includes materials concerning the Panama Canal and Indiana University.

SC65. 11 boxes (1913–1917), folder list. Presidential papers, chiefly correspondence with departments, arranged by subject or department. Includes materials relating to botany, chemistry, engineering, mathematics, mineralogy, physics, and other scientific fields, and to the American Association for the Advancement of Science, California Academy of Sciences, and American Society of Mechanical Engineers.

Campbell, Douglas Houghton (1859–1953), professor of botany, Stanford.

3 boxes (1888–1953). Correspondence, records of the biology department, class rolls, and some printed works by Campbell.

Chandler, Loren Roscoe (1895–), professor of surgery and dean of the Medical School, Stanford.

0900. Oral history.

Clark, Esther Bridgman, pediatrician, Palo Alto.

0905. Oral history.

Cutter, Lawrence (1877–1965), assistant professor of mechanical engineering, Stanford.

6 volumes (1902–1907), volume inventory. Stanford student notebooks and lecture outlines.

Earthquake (1906) collection.

1 linear foot (1906–1979). Eyewitness accounts, several by Stanford scientists, of the 1906 earthquake; clippings about the quake and its aftermath.

Ehrlich, Paul R., and Anne H. Ehrlich.

SC223. 44 linear feet (1961–1983). Videotapes of television appearances, professional correspondence and subject files, speeches, articles, tapes, films, interviews, and correspondence with the general public and with publishers, relating to his work as professor and her work as research associate in biological sciences at Stanford. Includes materials concerning Rocky Mountain Biological Laboratory, Jasper Ridge Biological Preserve, and Zero Population Growth.

Farnsworth, Paul Randolf, professor of psychology, Stanford.

Tape 1:19. Oral history, primarily concerning early days of the Department of Psychology.

Findlay, Alexander, professor of physical chemistry, University of Aberdeen.

> SC202:1:10. 1 folder. Diary kept during a visiting professorship at Stanford (1924–1925); news clippings.

Forsythe, George (1917–1972), professor of mathematics and computer science, first director of the Computation Center, Stanford.

> 34.5 linear feet (1938–1979), folder list. Professional correspondence, notes for lectures and publications, committee records, publications; plus materials relating to Alexandra Forsythe's interest in secondary-school instruction in computer science.

Fuchs, Henry O. (1907–), professor of mechanical engineering, Stanford.

> SC101. 6 boxes (1955–1974), restricted, folder list. Primarily course notes for mechanical engineering courses, case histories prepared by Fuchs, and photographs.

Geological Club of Stanford University.

> 1 v. (1893–1899). Minute book.

Geological Society of American Universities, Stanford Chapter.

> 2 v. (1901–1946). Records, minutes, memberships, Society constitution.

Gilbert, Charles Henry (1859–1928), professor of zoology, Stanford.

> 2 boxes (1880–1927). Notes, field notebooks, and journals containing notes and clippings about various species, particularly salmon.

Gorman, William H.

> 10 v. (1921–1922). Laboratory notes for Stanford courses in mechanical and electrical engineering, compiled by Gorman while an undergraduate.

Hansen, William Webster (1909–1949), professor of physics and director of the Microwave Laboratory, Stanford.

18 boxes (1925–1974), folder list and partial name index. Correspondence; manuscripts of writings; research notebooks; bibliographic and biographical information; his account of the development of the klystron and reports on the first linear accelerator at Stanford; class, lecture, and research notes. Includes materials concerning Sperry Rand Corporation, Varian Associates, and the MIT Radiation Laboratory.

3 v. Mimeographed "Notes on microwaves," based upon lectures at MIT Radiation Laboratory, and prepared by S. Sealy and E. C. Pollard.

Hoover, William, professor of electrical engineering and director of the Ryan High Voltage Laboratory, Stanford.

0900. Oral history.

Jacobsen, Lydik S. (?–1976), professor of mechanical engineering and director of the Earthquake Research Laboratory, Stanford.

0.5 linear feet (1948–1958). Miscellaneous correspondence and other papers relating to his research in seismology and structural dynamics.

Jahns, Richard H., professor of geology and applied earth sciences and dean of the School of Earth Sciences, Stanford.

0900. Oral history, including discussion of the U.S. Geological Survey.

Jenkins, Oliver Peebles (?–1935), professor and head of the department of physiology and histology, Hopkins Marine Station.

1 box (1888–1927). Research notes (1888, 1904); correspondence; unpublished and published writings, including addresses, essays, and pamphlets; accounts of meetings and field trips.

Johnson, Francis (1901–1960), professor of English, Stanford.

SC286. 6 linear feet (1932–1959), folder list; advance notice required. Biographical material, correspondence, research notes, programs of scholarly meetings, articles, and notes, some concerning astronomy and the history of science.

Jordan, David Starr (1851–1931), university president, Stanford.

400 boxes (1861–1964), finding aid, also available in a microfilm edition. Correspondence on education, international relations, eugenics, ichthyology, Stanford University; diaries and journals; lecture notebooks; published and unpublished writings; financial papers; biographical and genealogical materials; scrapbooks, clippings, and printed material.

SCM 015. 1 folder (1901–1902). Correspondence pertaining to the unsuccessful effort to obtain the Koenig Tonometer for Stanford.

PC8. 3 boxes. Photographs (negatives) taken by Jordan while serving on the Pribilof Islands Fur Seal Commission (summer 1897). Includes views of fur seal rookeries on St. Paul Island and St. George Island, Alaska.

Jordan, Eric Knight (1903–1926), youngest son of David Starr Jordan.

2 boxes (1909–1945). Correspondence regarding his conchological research, untimely death, and research fellowship founded in his name; first volume of his manuscript "Shells of California," and bibliography of his published papers.

Kaplan, Henry (1918–1984), professor of radiology, Stanford.

100–150 linear feet, unprocessed. Personal correspondence, administrative files, card catalog of reprint collection with notes and comments, research notebooks, reading and research notes, speeches, slides, etc., including materials concerning the Cancer Biology Research Laboratory.

Keen, Myra, curator of the Malacology Collection, Stanford.

0900. Oral history.

Kirkpatrick, Paul Harmon (1894–   ), professor of physics, Stanford.

83-013. Photocopy of autobiography completed in 1971, with author's bibliography to 1970.

Knuth, Donald (1938–), professor of computer science, Stanford.

14.5 linear feet (1962–1982). Materials pertaining to his *The art of computer programming*, including original manuscript, galley proofs, comments, and final version; materials regarding "Tex," a computer-based phototypesetting system.

Kompfner, Rudolf (1909–1977), professor of applied physics, Stanford.

SC194. 5 linear feet (1937–1980), folder list. General correspondence files, diaries (1937–1941), notebooks and notes (1937–1977), articles, reports (1937–1980), awards, and drawings; research notebooks regarding projects at Bell Laboratories. Includes materials concerning optics, traveling-wave tubes, Clarendon Laboratory, and the University of Birmingham.

Kreisel, Georg[e] (1923–), professor of logic and mathematics, Stanford.

SC136. 21 boxes (1957–1978), finding aid, restricted. Professional correspondence, lecture and seminar notes, manuscripts, and technical reports.

4 linear feet (1949–1981). Correspondence, notes, and memoranda from Kreisel to Jean van Heijenoort; some additional correspondence. Concerns symbolic logic.

Loewner, Charles (1893–1968), professor of mathematics, Stanford.

6 boxes (1923–1966), folder list. Lecture notes, reprints, some correspondence in German and English. Includes materials concerning complex analysis and differential geometry.

Luck, James Murray (1899–), professor of biochemistry, Stanford.

5.5 linear feet (1930–1979), folder list. Primarily correspondence and subject files relating to professional activities, the chemistry department, and his position as science attaché at American embassies in London, Stockholm, and Bern.

Marx, Charles David (1858–1940), professor of civil engineering, Stanford.

11 boxes (1909–1933). Correspondence, drawings, reports, and materials relating to consulting contracts.

Marx, Guido H. (1871–1949), professor of mechanical engineering, Stanford.

10 boxes (1897–1949), folder list. Correspondence; clippings; Marx's unpublished autobiography, pamphlets, and other publications relating to his professional, campus, and reform activities as a founder of the American Association of University Professors.

McFarland, Frank Mace (1869–1951), professor of histology, Stanford; co-director of Hopkins Marine Station.

1 box (1893–1943). Correspondence and writings, together with sketchbooks by Mrs. McFarland.

Moore, Arthur R., professor of biology, Stanford.

SC188. 1.5 linear feet (1905–1961), finding aid. Personal correspondence, class and laboratory notebooks, typed manuscripts, and rough drafts of articles, some concerning work at Hopkins Marine Station.

Moore, George J.

Papers, including materials concerning Hopkins Marine Station.

Morris, Samuel Brooks (1890–1962), professor and chairman of civil engineering, Stanford.

20 boxes (1936–1940). Primarily correspondence, notes, reports, and maps dealing with Morris' duties as water consultant for the National Resources Committee.

Mosher, Clelia Duel (1863–1940), physician, professor of personal hygiene, Stanford.

4 linear feet (1886–1938), finding aid. Writings, diaries, biographical and genealogical material, Mosher's survey of sexual practices of collegiate women, correspondence, and publications.

Park, Charles F., Jr. (1903–), professor of geology and dean of the School of Earth Sciences, Stanford.

22 boxes (1950–1974). Correspondence, including that concerning geological organizations; materials dealing with mineral exploration, work as a mining consultant, and the School of Earth Sciences.

Peirce, George James (1868–1954), professor of botany and plant physiology, Stanford.

1 box (1905–1926). Correspondence.

Poppe, Nikolai N. (1897–).

17 phonorecords. Oral history interview concerning the Soviet Academy of Sciences.

Rambo, William R. (1916–), professor of electrical engineering and director of Stanford Electronics Laboratories.

60 boxes (1939–1974), restricted, folder list. Papers reflecting administrative responsibilities and research interest, primarily during his directorship of the Stanford Electronics Laboratories (1960–1972).

Richardson, George Mann.

Correspondence, photographs, and scrapbook concerning work in organic chemistry.

Ryan, Harris J. (1866–1934), professor of electrical engineering and first director of the High Voltage Laboratory, Stanford.

1 box (1908–1934). Lecture notes, correspondence (1930–1932) regarding a lecture tour of Japan, clippings, and bibliographical information.

Schawlow, Arthur L. (1921–), professor of physics, Stanford; Nobel laureate.

Videotape of "From maser to laser," a talk delivered at Lawrence Livermore Laboratory (1982).

Videotape of press conference upon receipt of the Nobel Prize (October 1981).

Schiff, Leonard I. (1915–1971), professor of physics, Stanford.

SC220. 25.5 linear feet (1948–1971), folder list. Professional correspondence; reports and committee papers relating to scientific and administrative matters, including relativity, nuclear physics, quantum theory, the National Aeronautics and Space Administration, U.S. Air Force, and National Academy of Sciences.

Sears, Robert Richardson (1908–   ), professor of psychology, Stanford.

6 linear feet (1942–1965), restricted, carton list. Professional correspondence; research data; drafts and reviews of his publications; speeches; and correspondence pertaining to the American Psychological Association; also interviews for the Kansas City thumb-sucking study.

Semino, Angelo F.

0.5 linear feet (1919–1921). Student notes for Stanford courses in mechanical and electrical engineering; history of the Industrial Engineering Corporation founded by four Stanford engineers in the 1930s.

Shockley, William B. (1910–   ), professor of electrical engineering, Stanford.

SC222. 60 linear feet (1860–1978), restricted, folder list. Cassette tapes of telephone conversations, correspondence files, family records, genealogical information, photographs, and scrapbooks. Includes material concerning genetics.

Siefert, Howard (1911–1977), professor of aeronautics and astronautics, Stanford; director of the Physical Sciences Laboratory, United Aircraft Corporation.

39.5 linear feet (1960–1977). Correspondence, photographs, clippings, and office files. Includes materials concerning the Jet Propulsion Laboratory and American Rocket Society.

Skilling, Hugh Hildreth (1905–), professor of electrical engineering, Stanford.

SC37:2. 0.5 linear feet (1954–1964). Correspondence, memoranda, goals, statements, agendas, and notes pertaining to the Undergraduate Administration Committee in the Department of Electrical Engineering.

Stanford University. Administrative records.

Primarily folder or carton lists. Institutional records, including administrative records, university publications, brochures, technical reports, dissertations, committee reports, annual reports, many with information about science, technology, and medicine at Stanford.

——. **Biology, Department of.**

SC256. 2 linear feet (1949–1970), folder list. Primarily office files concerning the Hopkins Marine Station, minutes of faculty meetings, a study of housing for the Station, and office files concerning the Division of Systematic Biology. Includes information on the Te Vega Program.

——. **Earth Sciences, School of.**

0.5 linear feet (1903–1968). Miscellaneous records, including photographs of students and faculty, a history of the school of mineral sciences, architectural drawings, and portrait prints.

——. **Electronics Laboratories.**

SC214. 90 linear feet (1948–1977), restricted. Technical reports, electronics research reviews, and research notebooks, including 3 volumes concerning early research on traveling-wave tubes.

——. **Engineering, School of.**

27 boxes (1915–1969). Administrative records from the office of the dean, including correspondence, reports, memoranda and minutes arranged topically; files on teaching appointments; annual budget and reports to the university president.

———. **Commission of Engineers.**
5 boxes (1906–1908). Papers of the commission responsible for reconstruction of Stanford after the 1906 earthquake: correspondence, daily journals, photographs, estimates of repairs.

———. **Geology, Department of.**
SC164. 25 boxes (1907–1973), restricted, finding aid. Records, including student files kept by the department (1907–1913, 1924–1967), some correspondence, and minutes of departmental meetings.

SC259. 0.25 linear feet, 2 volumes (1917–1963). Guest book (1917–1947); correspondence regarding the malacology collection and WPA projects in the department (1937–1943); biographical information concerning Hubert G. Schenk.

———. **(William W.) Hansen Laboratories of Physics.**
19 boxes (1947–1964), restricted, folder list. Contracts and proposals; correspondence and memoranda; records relating to the construction of Stanford Linear Accelerator Center, microwave physics, and many other projects.

———. **Hypnosis Laboratory.**
SC241. Papers (1957–1979).

———. **Interviews.**
B24–32. Collection of taped interviews with Stanford faculty, including Alan Schwettman, William Fairbanks, and Henry Kaplan.

———. **Mechanical Engineering, Department of.**
SC171. 8 boxes (1920–1944). Primarily administrative correspondence from the department's records.

———. **Medical Center.**
Audio digest, with topic list by case and tape. Concerning patients' symptoms, diagnosis techniques, diseases, and treatment.

SC42. Carton list. Monthly reports (1914–1935) and census reports (1935–1940).

3 boxes (1938–1966), partially restricted. Records, financial reports, floor plans, appraisals, legal documents, clippings, etc., concerning Palo Alto Hospital, Stanford-Palo Alto Hospital, and Stanford Medical School.

Mural quadrant at Urbaniborg, from Tycho Brahe, *Astronomiæ instauratæ mechanica* (1602). Copies in the Stanford University Libraries and The Bancroft Library.

———. **News and Publications Service.**

Oral history interviews in the Notable Teachers Series, including interviews with Felix Bloch, Arthur Schawlow, Frederick Terman, and Donald Knuth.

Tapes documenting various topics, including the Einstein project, the Teller-Drell debate, acoustic microscopy, the Committee on Research, etc.

———. **Physical Sciences Program.**

10 boxes (1931–1973). Correspondence, budgets, and course materials from this program, primarily for 1960–1969.

———. **Physics, Department of.**

0.25 linear feet (1937–1947). Correspondence and news clippings about klystron research (1937–1945); correspondence and memoranda regarding department budget and faculty matters.

———. **President, Office of the.**

1 folder (1906–1913). Correspondence relating to the absorption of Cooper Medical College and Lane Hospital, San Francisco, by Stanford University.

———. **Psychology, Department of.**

SC277. 5.5 linear feet. Case files for mother-father interviews conducted in 1958, including interviews, related data, and impressions.

———. **Ryan High Voltage Laboratory.**

1 volume (1920–1931). Guest book kept by the Laboratory.

———. **Solar Observatory.**

B7. Tape recording of the dedication ceremonies (26 April 1974).

Stephen, Mark, instructor in communications, Stanford.

12 linear feet. Research material for his book, *Three Mile Island* (New York, 1980), including cassette tapes, interview transcripts, correspondence, and reports.

Stolz, Lois, professor of psychology, Stanford.

    0900. Oral history.

Swain, Robert Eckles (1875–1961), professor of chemistry and acting university president, Stanford.

    1 linear foot (1916–1938), folder list. Correspondence as department chairman and pertaining to the honor system in the department.

    Draft, typescripts, and notes for his address at the Stanford memorial service for David Starr Jordan (1932).

Terman, Frederick Emmons (1900–1982), professor of electrical engineering, dean of the School of Engineering, provost and vice-president, Stanford.

    220 boxes, finding aid. Correspondence, memoranda, minutes, patents, clippings, reports, speeches, publications, and other materials concerning Stanford, electronics in the Bay Area, radar research during World War II, etc.

    Transcript of Bancroft Library oral history. Includes discussion of electronics, Harvard Radio Research Laboratory, National Defense Research Committee (NDRC), Hewlett-Packard, and Silicon Valley.

    Tapes 1:8,15, 5:15. Tapes of statements made upon receipt of Stanford Associates' "Uncommon Man" Award (1978), at a Stanford dinner in his honor (1977), and at his Stanford memorial service (1983).

Terman, Lewis M. (1877–1956), professor of psychology, Stanford.

    66 boxes (1910–1959), restricted, folder list. Correspondence and data regarding Terman's psychological studies and intelligence tests; professional correspondence. Includes materials concerning the International Commission on Eugenics.

George Vanderbilt Foundation.

    1 box (1959–1967), restricted. Records of the Foundation and correspondence regarding its dissolution after Vanderbilt's death in 1961; agreements between the Foundation and Stanford University. Includes materials concerning ichthyology and the South Pacific.

Webster, David Locke (1888–1976), professor of physics, Stanford.

21 linear feet (1915–1973), carton list. Correspondence relating to research and the physics department; research notebooks; glass plates, photographs, and maps. Includes materials pertaining to klystrons and research on the Compton effect.

11.5 linear feet (1914–1964), folder list. Course material for his physics classes at Stanford, notes and notebooks, correspondence, subject files on Army research projects, departmental budget material, research data, photographs of apparatus, manuscripts and reprints of articles. Includes material dealing with Harvard, MIT, and the University of Michigan.

Weymouth, Frank Walter (1885–1963), professor of physiology, Stanford.

3 boxes (1907–1968). Correspondence, lecture notes, photographs and drawings of equipment, research notes, and manuscripts of writings.

Whitaker, Douglas Merritt (1904–1973), professor of biology, dean, acting vice-president, and provost, Stanford; chief biologist for the Bikini Atoll tests.

6 boxes (1926–1963). Speeches, articles, slides used in a lecture on evolution, some correspondence and lecture notes, and material concerning the Bikini Atoll bomb tests.

Wiggins, Ira Loren (1899–   ), professor of botany, director of the Natural History Museum and of Dudley Herbarium, Stanford.

0900. Oral history.

11 boxes (1933–1957). Primarily correspondence with colleagues; some manuscript material relating to publications.

Wilbur, Dwight, clinical professor of medicine, Stanford.

0900. Oral history.

Wilbur, Ray Lyman (1875–1949), professor of physiology and medicine, dean of Cooper Medical College, university president and chancellor, Stanford; Secretary of the Interior under Herbert Hoover.

    520 boxes (1914–1951), folder list. Correspondence; scrapbooks; publications; speeches; travel notes; rough drafts of "memoirs"; and subject files on institutions, organizations, friends and associates, conferences, and such topics as antivivisection.

Zimbardo, Phillip, professor of psychology, Stanford.

    SC180. 3 linear feet. Completed questionnaires from a study of shyness in various cultures.

St.02

Manuscripts Division
Department of Special Collections
Cecil Green Library
Stanford University
Stanford, CA 94305
(415) 497-4054

Berillon, Edgar (1854–1948), French physician.

83-028. Correspondence with physicians, neurologists, and psychiatrists on hypnosis, psychiatry, and medicine, including letters from the chemist Marcelin Berthelot.

Brasch family.

M176. 1 linear foot (1891–1920), finding aid. Primarily letters written by the children of Otto and Caroline Brasch to their brother, Frederick Brasch (see below). Includes accounts of the San Francisco earthquake.

Brasch, Frederick E. (1875–1967), history of science bibliographer; chief of Scientific Collections at the Library of Congress.

M275. 104 boxes, finding aid. Manuscripts written or collected by Brasch concerning the history of science, especially science in colonial America, and related subjects. Includes correspondence of T. J. J. See, Albert Einstein, and others.

Elwell, Cyril F. (1885–1963).

M49. 1.5 linear feet (1941–1961), folder list. Autobiography, notes, correspondence, clippings, speeches, magazine articles about Elwell, contracts, financial and engineering papers. Includes material on Hewlett-Packard and the Linear Electron Accelerator.

Exactus Photo-Film Corporation, Palo Alto.

Bo55. 1.5 linear feet (1914–1918). Records of the first documentary film company in the U.S., including records pertaining to inventions by T. K. Peters; patents; promotional brochures.

Fauche-Borel, Louis (1762–1829).

84-B61683. Unpublished manuscript, *Philosophie ayant rapport à la physiologie*... (1823).

Franklin, Benjamin (1706–1790).

84-040. 1 folder. Contemporary secretarial copy of Franklin's letter to John Lining (1759), regarding observations on electricity and other natural phenomena.

Heintz, Ralph M. (1892–1980), inventor and entrepreneur.

3 cartons. Tracings and sketches for a variety of mechanical and electrical devices, including a radar-hydraulic system, engines, refrigerators, etc.

Hopkins transportation collection, part V.

M234. 12 linear feet (1816–1942), folder list. Materials collected by Timothy Hopkins, including papers relating to the development of rail and air transportation, photographs of early vehicles, and the aviation papers of Stanley H. Page. Part of a larger collection pertaining to the transportation industry, especially railroads.

Hurst, George R.

Manuscripts, plans, blueprints, photographs, scrapbooks, catalogs, and other materials pertaining to the history of gold mining and dredging technology. Transferred from Jackson Library, Stanford.

Hyde, William Birelie (1842–?), engineer, Southern Pacific Railroad.

Ms267. 1 linear foot (1866–1877), finding aid, restricted. Correspondence relating to railroad engineering and telegraphy, including the Yosemite Survey and the Russian-American Telegraph Survey; Western Union telegraph message books (1873).

Jackling, Daniel C. (1869–1956), president of several mining companies.

M93. 18 linear feet (1911–1956), folder list, restricted. Private and personal files of correspondence, photographs and printed matter, ledgers and accounts concerning mining companies. Includes materials relating to Missouri School of Mines and professional organizations for mining and metallurgy.

Lewis, James.
M108. 5 boxes (1868–1873), box list. Correspondence with Annie Lawe regarding her shell collection, the first shell collection acquired by Stanford University (ca. 1891).

Medieval manuscripts study collection.
M299. (11th century–1582), item list. Manuscripts and fragments, including texts on medicine, philosophy, and meteorology.

Miscelanea [sic].
1 v. (ca. 1625). Anonymous manuscript in Italian, collecting texts on astronomy, mathematics, etc. Includes *De institutionibus arithmeticis, De sphera mundi, De praxi astronomica*.

Isaac Newton collection.
M132. 2.5 linear feet (1684–1949), finding aid. Correspondence, drafts, reports, notes, scientific manuscripts, lectures, tables, diagrams, and printed pamphlets by and about Newton and his influence.

Haskell Norman collection.
M336. (1752–1883), item list. Miscellaneous manuscripts, including several by 18th- and 19th-century scientists, including d'Alembert and Vaucanson, and anonymous manuscripts on mathematical topics.

Notebook of a medical student.
M337. 409 l. (18th century). Holograph notebook by an unknown medical student at the University of Montpellier. Includes transcript (copied from earler notes) of a discourse on pathology by Pierre Chirac, who taught medicine at Montpellier between 1687 and 1706.

Oldroyd, Mrs. Ida Shephard.
M197. 1 v. (1890–1927). Correspondence with William H. Dall about shell collecting and conchology.

Pacific Geographic Society.

3 linear feet (1927–1940), folder list; 1 day's notice required. Society records, including correspondence, minutes, reports, photographs, and pamphlets, primarily for 1933–1939.

Peckham, Stephen Farnum, petroleum engineer, California Petroleum Company.

M387. 0.25 linear feet (1865–1881), folder list. Journal and correspondence, geological field notes on Eastern oil fields, and notes on Eastern refineries.

Ricketts, Edward F. (1897–1948), president (?) of Pacific Biological Laboratories.

M291. 2 linear feet (ca. 1937–1979), finding aid. Letters; notes on intertidal marine life; printed articles; manuscript notes of unpublished biology articles and of his *Between Pacific tides*, including correspondence regarding the book and notes for revisions and additions; field notes.

Ritter, [August] Heinrich (1791–1869), professor of philosophy.

84–B61703. 1 box (1831–1832). Lecture notes from Ritter's course on psychology, signed on title by E. Burthmann, a student.

Scientists and science collection.

M133. 1.5 linear feet (1870–1948), folder list. Single items and small collections of material by or about scientists and their work, including letters, clippings, notes, diagrams, drafts of articles, speeches, and a journal. Concerns medicine, physics, geology, aerodynamics, etc.

Seale, Alvin (1871–1967), superintendent of Steinhart Aquarium, California Academy of Sciences.

M172. 2.5 linear feet (1901–1940), volume list. Collections of diaries, primarily containing biological (especially ichthyological and ornithological) observations, as well as accounts of his work at the Steinhart Aquarium and other posts. Includes materials concerning fisheries, Bishop Museum, and the Harvard Museum of Comparative Zoology.

See, Thomas Jefferson Jackson (1866–1962), astronomer and physicist.

3 boxes, folder list. Manuscript versions of his "New theory of the ether"; other manuscripts and notes on mathematics, physics, and astronomy; notebook of "Lowell Observatory accounts"; some correspondence and personal papers. See also St.02 Brasch.

Simplified spelling collection.

1 box (1916–1920), folder list. Papers from the movement for simplified spelling in the American Institute of Mining Engineers.

Strauss, Joseph, engineer.

Mi163. 20 linear feet (1905–1933). Collection of 93 sets of original bridge plans by Strauss.

Sutro, Adolphe.

JL (Jackson Library) 4. 1 linear foot (1869–1899), folder list. Papers dealing with the construction and engineering of the Sutro Tunnel, with some additional correspondence concerning instruments and inventions.

United States of America vs. American Telephone and Telegraph, et al.

M396. 3 linear feet (1974–1982). Photocopies of typescript documents and testimony presented during the antitrust case. Includes testimony by Arthur Schawlow, David Packard, Charles Townes, and others concerning Bell Laboratories Research Division, lasers, and microelectronics.

Above: part of the linear accelerator; below: William Hansen and some of the Mark I crew in the Stanford Quadrangle. Stanford University Libraries.

St.03

Archives
Hoover Institution of War, Revolution, and Peace
Stanford, CA 94305
(415) 497-3563

Agnew, Harold Melvin (1921–), physicist, Los Alamos Scientific Laboratory (1943–1946).

1 reel (1945). Film of the explosion of atomic bombs over Hiroshima and Nagasaki.

Agriculture collection.

1 folder (1889–1939). Collection of pamphlets and bulletins (in English, French, and German) relating to international agricultural research, production, and expositions.

Beggs, James Montgomery (1926–), associate administrator of the National Aeronautics and Space Administration; Undersecretary of Transportation.

61 boxes (1959–1981). Speeches and writings, correspondence, memoranda, reports, studies, notes, printed matter, and photographs relating to transportation, space exploration, and other topics.

Billings, Bruce Hadley, physicist; Special Assistant for Science and Technology to the U.S. Ambassador in Taiwan (1968–1973).

1 box. Mimeographed "Study of the role of science and technology in Taiwan," with special attention to issues of economic development.

Blackwelder, Richard Eliot (1909–).

79005-10. 2 scrapbooks (1945–1946). Clippings relating to the development of the atomic bomb.

Butler, Charles Terry (1884–?), physician.

1 v. Photocopy of unpublished typescript autobiography, "A civilian in uniform," pertaining to medical education in the United States (1912–1916) and to the U.S. Medical Corps in World War I.

Cranberg, Lawrence (1917–   ), physicist.

260 p. (1980). Photocopy of unpublished typescript, "Detour to dystopia: A century of pseudo-scientific socialism," relating to the claim of scientific status for the works of Karl Marx and Friedrich Engels.

Darling, William Lafayette (1856–1938), civil engineer; member of the U.S. Advisory Commission of Railway Experts to Russia.

1 v. Diary relating to the Russian railway system (May-December 1917).

Deutsche Forschungsgemeinschaft.

6 boxes (1834–1936). Correspondence files relating to grant applications for research.

Elliott, John Elbert (1887–1980), oil geologist, engineer, and entrepreneur.

62 boxes (1902–1980). Letters, photographs, clippings, printed material, and memorabilia, primarily relating to the oil industry, petroleum engineering, geology, and personal matters.

Emerson, George H., commander of the Russian Railway Service Corps.

½ box (1918–1919). Correspondence, reports, maps, photographs, and clippings relating in part to the activities of the Russian Railway Service Corps.

Farquhar, Percival (1864–1953), engineer.

4 boxes (1922–1928). Correspondence, memoranda, and reports pertaining to negotiations regarding Russian iron ore and steel resources and to the work of American engineers in the Soviet Union.

Foss, F. F., engineer in Russia.

0.5 linear feet (1890–1917). Photographs and memorabilia relating to engineering and to the development of industry in prerevolutionary Russia.

Gilinsky, Victor (1934–), physicist; member of the U.S. Nuclear Regulatory Commission.

63 boxes (1972–1982). Reports, memoranda, transcripts of hearings and telephone conversations, correspondence, and printed matter, relating to nuclear power plants in the U.S. and especially to the Three Mile Island accident.

Gilman, G. W., engineer, Bell Laboratories.

Ts Japan G487. 1 folder (1932–1934). Typewritten memoranda relating to the organization and equipment of an international radiotelephone communications system in Japan.

Gorton, Willard L. (1881–?), civil engineer; consultant to the Soviet government in Turkestan.

2 boxes, 313 prints, 172 negatives (1927–1932), folder list. Correspondence, reports, clippings, and photographs relating to reclamation and irrigation projects in Turkestan.

Gunn, Selskar M., officer of the Rockefeller Foundation.

Vw China G976. 1 v. (1934). Mimeographed report, "China and the Rockefeller Foundation," pertaining to educational, scientific, and cultural assistance activities of the Foundation in China.

Hiroshima-Nagasaki Publishing Committee.

2 reels (1982). Compilation of film clips depicting the destruction and the condition of survivors of the atomic bomb blasts.

Hoover, Theodore Jesse (1871–1955), dean of the School of Engineering, Stanford.

1 box, 1 reel (1939), restricted. Photocopy and film of typewritten memoirs entitled "Memoranda: Being a statement by an engineer," relating to engineering and to the Hoover family.

Hoskin, Harry L. (1887–), officer in the Russian Railway Service Corps (1917–1920).

½ box (1917–1973). Correspondence, clippings, reports, affidavits, court proceedings, and photographs pertaining to the activities of and disputes about the Corps.

Hutchinson, Lincoln, president of the American Relief Administration Association.

2 boxes (1923–1935), folder list. Correspondence, writings, and reports relating in part to technical assistance provided by American engineers in the Soviet Union.

Interessengemeinschaft Farbenindustrie Aktiengesellschaft. Propaganda-Abteilung.

Ts Germany I42. 1 folder (1932–1942). Mimeographed bulletins (in German) relating to conditions in the chemical industry in Germany.

Intergovernmental Oceanographic Commission.

1 box (1962–1964). Reports, bulletins, and minutes of meetings, relating to the promotion of international oceanographic research under the auspices of UNESCO.

Johnson, Benjamin O. (1878–?), engineer.

1 box (1917–1923), folder list. Correspondence, reports, memoranda, diplomatic dispatches and instructions, and printed material concerning the Russian Railway Service Corps in Siberia, the Inter-Allied Technical Board, and related matters.

Kantor, Harry Simkha (1903–1980), analyst for the U.S. Department of Labor.

32 boxes (1938–1976). Correspondence, writings, reports, memoranda, notes, legislation, and printed matter. Topics covered include the encouragement of inventors in the U.S.

Kennedy, Richard Thomas (1919–), member of the U.S. Nuclear Regulatory Commission.

92 boxes (1975–1980). Speeches, correspondence, memoranda, reports, transcripts of hearings, and printed material relating to nuclear power plants in the U.S. and especially to the Three Mile Island accident.

Kirkpatrick, Paul Harmon (1894–), professor of physics, Stanford.

> 1 box (1955–1958). Correspondence, memoranda, and reports relating to security clearance procedures for education advisors in the U.S. International Cooperation Administration and to education in the Philippines, especially the teaching of physics at the University of the Philippines.

Lademan, Joseph V. (1899–), captain, U.S. Navy.

> ½ box (1942–1969). Memoranda, writings, reports, printed matter, clippings, and photographs relating in part to Bikini Atoll bomb tests (July 1946).

LeBaron, Robert, Deputy Secretary of Defense for Atomic Energy.

> 6 boxes (1949–1981). Biographical sketches, reminiscences, speeches, correspondence, appointment books, memoranda, memorabilia, and photographs relating to the development of atomic weapons and to peaceful uses of atomic energy.

London, Ivan D., American psychologist.

> 90 boxes (1943–1982), restricted. Correspondence, writings, questionnaires, interview transcripts, notes, reports, memoranda, and printed matter, partly concerning the study of science, especially psychology, in the Soviet Union.

Luck, James Murray (1899–), professor of biochemistry, Stanford.

> 1 box (1941–1960). Reports on food supply in the U.S., Canada, and Great Britain (1941), and on the American scientific exchange mission to the Soviet Union (1960); photographs.

Metzger, H. Peter, science editor for the *Rocky Mountain news*; president of the Colorado Committee for Environmental Information.

> 30 boxes (1945–1981). Writings, clippings, reports, studies, printed matter, and correspondence relating to nuclear energy, environmental problems, and the impact of science and technology on public policy.

Miller, Oliver W., colonel, U.S. Air Force.

1 box (1943–1972). News releases, radio messages, printed matter, photographs, clippings, maps, etc., relating to the development of radar and its applications in warfare.

Nederlandsche Artsenkamer.

Ts Netherlands A792. 1 folder (1941–1943). Mimeographed notices concerning regulations pertaining to physicians in the Netherlands during the German occupation. Includes materials relating to the Nederlandsche Vereenigung van Ziekenfordsarten.

New South Wales Medical Board.

1 folder (1932). Memoranda concerning permission for foreign or foreign-trained physicians to practice medicine in New South Wales and the exclusion of Germans and Austrians.

Packard, David, co-founder of Hewlett-Packard.

Oral history.

Pash, Boris T., colonel, U. S. Army.

1 box (1918–1963). Correspondence, memoranda, reports, orders, writings, photograph albums, and printed matter relating in part to the Alsos mission to determine the status of German nuclear development (1944–1945).

Pictorial miscellany.

12 envelopes, 1 box of transparencies (ca. 1900–1980). Includes photographs of atomic weapon testing.

Poncelet, Eugene F., metallurgical engineer.

1 folder. Memoirs relating to his life in Belgium before and after World War I, and to engineering in Canada and the U.S. after his emigration.

Rabbitt, James Aloysius (1877–1969), mining engineer.

61 boxes (1895–1969), finding aid. Correspondence, memoirs, lectures, reports, surveys, patents, clippings, sketches, and photographs pertaining to economic, scientific, and technological

developments in mining and metallurgy in China, Japan, and the Far East.

Ray, Dixy Lee (1914–), chairwoman of the U.S. Atomic Energy Commission.

178 boxes plus additional materials (1973–1982), restricted. Correspondence, speeches, reports, studies, printed matter, and audio-visual materials, relating in large part to nuclear energy.

Rhodes, Fred Burnett, Jr. (1913–).

22 boxes. Includes materials relating to the Manhattan Engineer District.

Ruark, Arthur Edward (1899–1979), senior associate director of research for the U.S. Atomic Energy Commission.

61 boxes (1885–1979). Writings, technical reports, preprints, lectures, correspondence, notes, and printed matter relating to physics.

Russian subject collection. Formerly the "American engineers in Russia" collection.

4 boxes (1927–1933), finding aid. Correspondence, writings, articles, and answers to questions relating to working conditions and to Soviet administration of American engineers.

Sams, Crawford Fountain (1902–), brigadier general, U.S. Army Medical Corps.

79066-10. 16 boxes (1923–1939). Correspondence, orders, speeches and writings, research data, printed matter, certificates, and photographs relating to military medical activities and to scientific research on the biological effects of radiation.

Starr, Clarence T., American mining engineer in the Soviet Union (1928–1931).

1 box (1923–1941). Correspondence, writings, notes, transcripts of testimony, and printed matter relating in part to coal mining and engineering in the Soviet Union.

Stepanov, Afansiy Ivanovich, Soviet engineer.

2 boxes (1956–1961). Correspondence, legal documents, memoranda, reports, engineering diagrams, and photographs

relating to proposals for technological innovations in Soviet industry and to Soviet factory management, etc.

Stevens, John Frank (1853–1943), civil engineer; chairman of the U.S. Advisory Commission of Railway Experts to Russia (1917).

½ box (1917–1931). Memoirs and correspondence relating in part to railroads and engineering during the Russian Revolution and to the Inter-Allied Railway Commission.

Stuart, Charles Edward (1881–1943), engineer.

4 boxes (1917–1942), folder list. Writings, correspondence, memoranda, reports, clippings, and motion picture film relating in part to the American fuel industry and to coal mining and engineering in the U.S. and the Soviet Union.

Teller, Edward (1908–   ), professor of physics, UCB.

17 boxes (1940–1975). Speeches and writings, reports, magazine and newspaper clippings, correspondence, photographs, motion picture films, awards, and other materials relating to physics, energy research, and security issues.

Tokyo Daigaku statement.

Ts Nuclear Warfare J9 T64. 1 folder. Printed statement by members of the College of General Education of Tokyo University (11 July 1957) relating to the dangers of radioactive fallout due to testing of nuclear weapons.

U.S. Advisory Commission of Railway Experts to Russia.

½ box (1931–1936). Correspondence of former members of the Commission relating in part to its activities in Siberia in 1917–1923.

Weakley, Charles Enright (1906–1972), assistant administrator of the National Aeronautics and Space Administration.

4 boxes (1945–1972). Correspondence, orders, drafts of speeches, printed matter, photographs, and sound recordings relating to U.S. anti-submarine force operations and to the activities of NASA.

Witkin, Zara, American construction engineer in Russia (1932–1939).

½ box (1933–1937). Writings and correspondence relating to Witkin, including his memoirs and a biographical sketch by Eugene Lyons.

Young, Millard C., brigadier general, U.S. Air Force.

2 boxes (1932–1980). Orders, personnel records, correspondence, speeches and writings, pamphlets and other printed matter, citations, and photographs relating especially to the development of the atomic bomb.

Yurchenko, Ivan.

25 boxes. Holograph manuscripts in Russian, dealing with philosophy, religion, the sciences, and other topics.

St.04

Stanford Linear Accelerator Center (SLAC)
Stanford, CA 94305

Director, Office of the.

Administrative and personnel records; records and office files of the director (ca. 1962–); restricted.

Scientific records.

Office files of the director, publications, and other records (ca. 1962–).

Department of Energy administrative records.

Financial, administrative, personnel, and other records (ca. 1962–); restricted. Some records have been transferred physically to the U.S. Federal Records Center in San Bruno, CA. For further information, write or call the Office of the Director at SLAC.

High-energy physics collection.

Comprehensive collection of preprints, reports, transcripts of talks, and informally published materials concerning high-energy and particle physics and related topics, including an archival collection of SLAC preprints and reports. Located in the SLAC Library, with bibliographic access provided through an online database (HEP) containing more than 100,000 records.

St.05

SRI International
Library and Research Information Services
333 Ravenswood Avenue
Menlo Park, CA 94025
(415) 326-6200

Holdings of the SRI Library include a collection of internal technical reports, contracts, contract proposals, and records relating to SRI research projects; project background files; and some project histories. Research areas include engineering, life sciences, physical sciences, weapons research, and energy and environment studies. Access to SRI materials is restricted. For specific information, write or call the director of Library and Research Information Services.

St.06

Lane Library
Medical Center
Stanford University
Stanford, CA 94305
(415) 497-6831

Manuscript holdings of Lane Library include records of the Stanford University Medical Center and of its predecessors, the Medical Department of the University of the Pacific and Cooper Medical College, as well as other manuscript collections pertaining to medicine and the life sciences. Arrangements should be made with the director for access to materials.

Abu-Hamid Muhammed ibn abi ibn Umar Najib-al-din-ul-Samarquandi, Kitab-i-Agdzieh we Ashribeh.

Mss Z286. 212 double p. Arabic manuscript, "Book of foods and drinks."

Al Ibrahimee, Kitab-i-Mudjerrabat min el Felâssefeh w'al Atibbâ.

Mss Z274. 95 p. Arabic manuscript, "Experiences, or tested prescriptions of philosophers and physicians."

Al Kourashi, Ash Shamil fis Senaat ut Tibbieh.

Mss Z276. 3 v. (13th century?), also available on microfilm. Fragment of an extensive treatise in Arabic on medicine.

Alvarez, Walter Clement (1884–).

R172.5 A49. Correspondence (1971–1976) with Clara S. Munson; related bibliographic reprints.

Mss H172.5H A47 1963. Typescript letter regarding Albert Abrams.

Mss 16. 7 cartons, unprocessed. Papers.

Mss H172.5H A48 1961. Recollections of Shad Bensley.

Typescript memoirs regarding Cooper Medical College by Luis F. Alvarez (class of 1887) and Walter C. Alvarez (class of 1905).

Bahr al Gauhar, Muhammad ibn Usef al Tab ul Harawi.

Z292. 237 l. (1813). Manuscript, "The sea of jewels" regarding the uses of medicinal plants, animals, and stones.

Barkan, Adolph (1845–1935), professor of medicine, Stanford.

Mss H154H B254. 5 boxes. Correspondence and papers.

Mss Q16H B25. 1 v. (1878–1910). Record of diseases of the eye.

Bettman, Jerome W.

Mss 7. Typescript, "Plan for regional courses to update the practicing specialist," by Bettman and others. Concerning ophthamology.

5300/2. "Vis a tergo: A history of the eye department" at Stanford University Medical Center.

Black, Joseph (1728–1799), professor of chemistry, University of Edinburgh.

C28H B62 1777. 4 v., also on microfilm. Notes by a student, titled "The substance of a course of chemical lectures as delivered by Dr. Black."

C28H B62. 8 v. (1773–1774?), also on microfilm. Notes by a student, titled "Lectures on chemistry delivered in the University of Edinburgh."

Bodonus, Carolus Antonius (?–1799).

Mss H128.7H B58. 2 v. Manuscript of *Botanica sev herbarvm*, with additional manuscript and printed material.

Bone and joint deformities and tumors.

Mss M684.H B71. 7 albums of photographs (ca. 1900) without captions or text.

Breit, Alfons.

Mss 11. Papers.

Brown, Adelaide (1868–1940).

Mss H710H B87. 1 box. Correspondence and papers.

Brown, Benjamin Boyer.

V125H B87 1830. 1 v. Manuscript titled "A medical, surgical, and pharmaceutical compendium in 6 pts., being an extraordinary compound, composed of ordinary simpels [sic]," Philadelphia.

Mss V126H B87 1830–1832. 1 v. Notes made by a student of medicine at the University of Pennsylvania.

Brown, Philip King (1869–1940).

Mss 1. 3 boxes (1933–1940). Correspondence, reports, publications, and other materials relating to medical practice and policy, especially in California.

California Society for the Promotion of Medical Research.

Mss H10H C16. Register of checks drawn.

Case of diabetes mellitus.

Mss P476H D5C3 1917. Record of the case of a 10-year-old child in the pre-insulin era, of interest because of treatments prescribed (starvation, whiskey, diet).

Celsian Association of the Medical College of the Pacific.

Mss H747.83H C39. 2 v. (ca. 1879–1882). Records, including the Association's constitution and minutes of meetings.

Chandler, Loren Roscoe (1895–   ), professor of surgery and dean of the Medical School, Stanford.

1 box (1942–1945). Correspondence as dean; catalog cards of his publications.

Cooper Medical College.

Mss H747.6H C7C8. Annual reports (1895–1905) of the president of the College, together with related manuscript material.

Mss H747H C7C7. 2 boxes (1889–1911). Class, attendance, and miscellaneous records; class programs.

Cooper Medical College (continued).

Commencement records (1883–1912).

Mss H747.55H C77. Minutes from meetings of the Committee on Clinics (1882–1897).

Mss H747H C7C81. 12 boxes. Miscellaneous correspondence.

Mss P47 C77 1883–. 18 v. (1883–1910). Records of diseases reported in the Children's Clinic.

11 v. (1887–1896). Records of clinical cases of eye, ear, nose, and throat disease.

Mss H747.6H C7. 1 folder, 5 v. Minutes of faculty meetings, in draft and final versions; related materials.

Mss H747H C7C82. 10 boxes. Financial documents.

1 folder. Founding records.

Mss N206 C77 1897–1910. 15 v. in 16. Records of gynecological cases; first 3 volumes wanting.

Mss H747.55H C78. 4 v. Minutes of the board of directors, articles of incorporation, and by-laws.

Mss J124H C77. 41 v. (1899–1933). Pathology records and letter copybooks.

Mss H747.8 C77. 4 v. (1905–1915). Clinic records.

Mss H747H C7C83. 1 box. Records, including clinical records and housing lists.

Mss L346 C77 1883–1911. 7 v. Records of the Neurological Clinic and of cases of "Nervous diseases."

Mss N30 C77 1895–1906. Records of genito-urinary disease cases.

Mss M34 C77 1883–1904. Records of surgical cases.

Mss L66 C78 1899–1907. 16 v. Records of the Medical Clinic.

Mss L66 C77 1883–1895. 8 v. "Medical diseases": records of the Medical Clinic.

Mss H747.8H C73. 1 v. (1883–1903). Registry of students and graduates.

Mss T72 C77 1902–1910. 2 v. Skin disease case records.

Mss H747.6H C7C5. Treasurer's and financial reports for 1899, 1903, 1905.

Mss H747H C7A4 1903. Membership list for the Alumni Association, compiled by its secretary-treasurer, Ray Lyman Wilbur.

Mss H747.85H C7. 2 v. Alumni Association records.

Cox, Alvin J. (1907–).

R747 S7C88 1980. 1 box. Typescript of lecture, titled "Stanford Medical School in San Francisco, the parent organization"; slides.

R747 S7C87 1979. Transcript of a lecture, titled "Stanford Medical School—the San Francisco years."

Crawford, Albert Cornelius (1869–1921).

Mss H154H C898. 2 v. Pharmacology notebooks concerning the pituitary gland.

Mss H154H C899. 2 boxes. Primarily research notebooks and reports pertaining to experiments.

Cushing, Clinton (1840–1904).

N231H C98 1889. Typewritten notes taken by Dr. Emmet Rixford for Cushing's gynecology lectures.

De Forest, Clara.

Mss W181H C1S8. Two copies of "historical data" concerning the Stanford School of Nursing.

Dickson, Ernest Charles (1881–1939).

Mss I1260 P21. 4 v. Unprocessed. Papers on botulism collected by Dickson.

Mss I1260 D53. 3 v. (1917–1921). Correspondence from California Packing Corporation, Canners' League, and National Canners' Association, concerning botulism.

Doctor's record books.

Mss 12. 3 v. (1889–1907). Anonymous records of cases treated.

Dolger, David.
Mss H747H S7D6. 11 l. (1975). Photocopy of typescript history, titled "Stanford School of Medicine fifty years ago, 1923 through 1925."

Doremus, R. Ogden.
Mss E71 D69 1865. 1 v. Manuscript of lectures on anatomy and physiology, with hand-drawn illustrations.

Drinker, Cecil Kent (1887–1956).
Mss F115 D76 1941. Lane Lectures about the lymphatic system, physiological and clinical considerations.

Duke-Elder, William Stewart.
Mss Q26H D87. Three letters (1942–1943) concerning surgery for cataracts.

Duncan, Charles H.
Mss V741 D92. Papers on autotherapy.

Eloesser, Leo (1881–?).
Mss H710H E49 and E491. 1 box (?–1974). Correspondence, reprints, curriculum vitae, list of publications, and photographs.

Faber, Harold Kniest (1884–1979).
Mss 5. 2 boxes (1911–1979). Papers, including Army Service papers, bibliography, Pediatrics Department history, materials relating to Faber Library, as well as personal and family letters of his father and his wife, Mary Faber, concerning her association with the Stanford Convalescent Home.

Fletcher, Horace. (1849–1919).
Mss H154H F61. 1 box. Correspondence with Dr. Borossini regarding "Fletcherizing" and other topics.

Gayet-ul-Beyan fil Tib.
Mss Z275. Translation (1775) into Turkish of an Arabic manuscript on medicine written ca. 1200.

Geiger, Jacob Casson (1885–).
Mss I11 B21 no. 127. 1 box (1922). "Epidemiology of botulism."

Ghigath-al-Din Ali al Husayn al'Isfah in, Dânish Namel.
Mss Z270. 118 double p. "The book of knowledge," Hindustani manuscript on natural history.

Gibbons, Henry. (1840–1911).
Mss 710H G44. 2 boxes. Miscellaneous papers, primarily incoming correspondence.

Gould, Agnes Safley.
Mss 15. 1 box (ca. 1906). Letters and manuscripts regarding her experiences as a student nurse in the Lane Hospital Training School for Nurses and the San Francisco Earthquake.

Gunn, Herbert.
Mss L862 G97. Typescript, "Amebic granulomata of the large bowel and their clinic resemblance to carcinoma," by Gunn and Nelson J. Howard.

Halls of Aesculapius.
Mss H747.36H S7. 1 box (ca. 1903). Plans, correspondence, etc., concerning the "Halls of Aesculapius," designed for Levi Cooper Lane and his wife, but never constructed.

Hanzlik, Paul John (1885–1951).
Mss D1512H H25 1908. 1 folder. Manuscript material concerning comparative anatomy of dogs.

Hebra, Ferdinand, Ritter von (1816–1880).
Mss T71H H43. "Ueber Hautkrankheiten."

Higgins, William M. (1826–?).
Mss 8. Letter (1890) to Dr. Henry Gibbons regarding Alice Higgins (1836–1890), the first woman graduate of Cooper Medical College.

Hospitals in California.
Mss I981H C15. 7 boxes. Correspondence and miscellaneous information concerning the history of hospitals in California.

Hunter, William Strobel.
Mss 2. 11.5 linear feet. Translations, critical notes, and correspondence regarding human locomotion.

Huntington, Thomas W.
Mss H708H H95 1906. Typescript of his Lane Lecture, "Dr. Levi C. Lane and academic hospitals."

Kâmrân Shirazi, Medjumat-us-Sanai.
112, 164 l. (1880). Persian manuscript, "A cabinet of arts," bound with "Tufut-ul-Mûlîk" a collection of tracts on medicine and folklore, by Yusufi.

Karabadin Dehelvi.
Mss Z272. 267 double p. Hindustani manuscript, "On remedies."

Lane, Levi Cooper (1833–1902), physician and medical educator; founder of the medical college that later became Cooper Medical College.
Mss H747H C7L4. 6 v. (1866–1873). Account books.

Mss M172.5H L26. Manuscripts of addresses delivered before medical colleges, including the Cooper Medical College and Lane Hospital; also lectures on such medical topics as aneurisms (1861–1895).

Mss L172.5 L29. Handwritten preliminary draft of the constitution of the San Francisco Medico-Chirurgical Society.

Mss M108H L26. Typescript, "Fractures of the bones which form the elbow-joint, and their treatment."

Mss I982H S2L3. 1 notebook. Notes for a history of Lane Hospital.

Mss H172.5H L28. 6 diaries.

Mss H747H S7S91. 4 folders (1886–1899). Manuscript copies of lectures in the Lane popular lecture series.

Mss H172.5H L29. 10 boxes, 2 photograph albums. Miscellaneous manuscript materials; photograph albums; notes from Esmarch's prize essay (1877).

Mss H708H L27 1896. 1 folder. Manuscript, "Progress in medicine; tribute to Michael Servetus."

Mss H172.5H L27. 1 v. (1877–1880). Record of patients, visits, diagnosis, and payments.

Lane Hospital, San Francisco.

Mss I982H S2L26. 1 v. (1894–1906). Minute book for meetings of the board of managers.

Mss I982H S2L27. Minutes of meetings of the Hospital Committee (1907–1911).

Lane Medical Library.

Mss H747.3H L2. 2 boxes. Miscellaneous correspondence, including some correspondence of Levi Cooper Lane.

Mss 19. 1 box (1913–1951). Miscellaneous correspondence files on medical and library matters.

3 v. (1895–1971). Record of borrowers; membership records.

Lee, Russel Van Arsdale (1895–).

Mss 4. 27 linear feet (1935–1976). Correspondence and assorted materials concerning aging, military interests, Medico Inc., and the Agency for International Development (?).

Leeches.

Mss D1091 H6L4. Anonymous manuscript, "The American medicinal leech."

Manson, Clara S.

Mss 13. 1 box (1927–1964). Primarily correspondence regarding the Lane Medical Library.

Medical College of the Pacific, San Francisco.

Mss H747 M43. 3 v. Minutes of faculty meetings.

Mss H747H M485. 1 pamphlet binder. Miscellaneous manuscript materials.

Medicinalia et chirurgica.

Mss H111H M495 1800. A collection of manuscript notes concerning medical subjects appearing in the *Reichsanzeiger* for 1800, with poetry and elogies added by D. Heinich Schmalzing.

Mir Muhammed, Ilmi Tib.
Mss Z277. 400 p. (1809). Persian treatise on materia medica, written in Talik.

Morton, William Thomas Green (1819–1868).
Mss M79H M67. 1 envelope. Manuscript and photographs concerning his use of ether.

Mosher, Clelia Duel (1863–1940).
Mss H708H M91. 16 v., unprocessed. Notes on various scientific, medical, and literary topics.

Moustafa ibn Halil us-Safiee, Kitab-ul-Edjsad.
Mss Z283. 201 l. (ca. 1840). Arabic treatise on alchemy.

Muhammad Akbar, Karabudin Kaderee.
Mss Z266. 205 l. (1830). Hindustani manuscript on remedies, materia medica.

Muhammad alad-din bin Hibat-Allah Adbzawari, Zubdat-u-Kawanin-i-Iladj.
Mss Z269. 116 p. (1717). Hindustani manuscript, "Treatise on remedies."

Muhammad Mu'min Husaini, Tufet-ul-Muminin.
Mss Z287. [528 l.] Persian manuscript, "Treatise on materia medica."

Muhammed Akber, Mudjerrebat-i-Akberee.
Mss Z273. 127 double p. Hindustani manuscript, "Treatise on the treatment of disease."

Munshi-Mazhar-al-Din.
Mss Z296. 231 l. (1815). Persian treatise on geography, with illustrations and maps.

Muzaffer bin Muhammed ul-Hvsny, Tibb-i-Shiffa'ee.
Mss Z271. 98 double p. Manuscript in Persian, "Remedies and their uses."

Narcylene collection.
Mss M86 N2Z1. Unprocessed collection of manuscripts.

Nedjibeddin es-Samerkandi, Karabidin-un-Nedjib es-Samarkandi.
Mss M284. 55 double p. Incomplete Arabic materia medica.

Ophüls, William (1871–1933).
Mss H710H O61. 2 boxes. Personal and professional correspondence and miscellaneous documents.

Mss J57 O61 1926. Manuscript of a statistical survey of 3000 autopsies.

Pamphlets on birth control.
Mss O856 P18. Pamphlets, together with related correspondence and manuscript materials.

Pathologia generalis et speciales.
Mss Z282. 29 l. (ca. 1840). Anonymous Arabic manuscript.

Plummer, Richard H. (1840–1899).
Mss H710H P73. 1 v. Lecture notes taken in 1864–65, with a diary for part of 1867.

Polio research.
Mss P496 P7S7 1947. 1 v. Newspaper clippings on isolation of polio virus, collected by Hubert S. Loring and Carlton E. Schwerdt of the Stanford School of Medicine.

Pont, Antonino Marcó del.
Mss L150H P81w 1918. Manuscript history of influenza epidemics.

Prentice, Arthur Dudley (1869–1959).
Mss 9. Two letters to Hans Barkan regarding Cooper Medical College graduates (1954); copies of obituary notices about Prentice.

Reed, Alfred Cummings (1884–1951).
Mss 6. 1 box (1916–1929). Correspondence from Ray Lyman Wilbur, William Ophüls, and others; notebook of lectures at the London School of Tropical Medicine.

Rixford, Emmet (1865–1938).

Mss H172.5H R62. 4 boxes. Miscellaneous manuscript materials.

Mss H172.5H R63. Portfolio of photographs of San Francisco County Hospital.

Rytand, David A.

5300/21. Letter titled "History of research in [the] Department of Medicine at Stanford," written in response to a request by A. McGehee Harvey.

San Francisco City and County Hospital.

Mss N206 S19 1904–1911. Records of City and County Hospital gynecological cases treated at Lane Hospital.

San Francisco County Medical Society.

Mss H10H S19. 1 box, 2 v. (1893–1902). Miscellaneous records and minutes of meetings.

San Francisco Maternity.

Mss O16H S1 S1–S3 and O129 S19. Minutes of meetings of the board of directors, records of members, scrapbook, and records of confinements and applications (ca. 1904–1921).

Sanitary surveys.

392 v., including indexes.

Sarif Han, Iladj-ul-Amradz.

Mss Z267. 562 double p. (1763?). Hindustani manuscript, "On diseases and remedies."

Segura, Ignacio Joseph, Spanish physician in Mexico.

Mss H128.7H S45 1745–1749. Unpublished (?) medical studies.

Mss L106H S45 1732. Manuscript, "Tractatus de febribus," pages wanting at end.

Shal mardan al Mustawf, Nuzhat Nameh-i-Alai.

Mss Z279. 151 p. Persian manuscript concerning natural history, geography, physics, and medicine.

Si hat äd-din bin Abd äl-Karim, Shifâ-ul-Maradh.

Mss Z280. 186 p. (1776). Persian manuscript, "The curing of disease," in verse.

Sinhalese ola.

Manuscript notebook (ca. 1800) of a medical student in Ceylon, concerning "Diseases of the eye" and their treatment. Written on prepared strips of Talipot palm leaves.

Also a Sinhalese herbalist or medical recipe ola written (ca. 1750) in Pali on the prepared leaf of the Talipot palm.

Sovary, Lilly.

Mss 17. 1 box, unprocessed. Papers.

Spielmann, Marion H. (1858–1948).

Mss E21H V58S8. Manuscript material, photographs, and print of a plate of Vesalius, used in writing the *Iconography of Andreas Vesalius* (1925).

Stanford Clinics Auxiliary and San Francisco Maternity.

Mss O16H S1S4. 3 boxes. Records, correspondence, reports, and other documents.

Stanford University. Medical Center.

Mss H747H S7S9. 1 box. Correspondence of the School of Medicine.

Mss H747H S7S94 1912. 1 folder. Letters, primarily from Hugh Goodfellow, attorney at law, concerning establishment of a fund to continue the Lane Medical Lectures.

Mss O129 S79 1912/30. 18 v. Gynecological Clinic specimen reports.

Mss H747H S7S94H 1959. "The first hundred years," typescript history of the Stanford University School of Medicine.

Mss H747H.S7S96.1912–14. Minutes of the Medical Division; report of the History Committee.

Mss H747H S7S92. Miscellaneous manuscript material from the School of Medicine.

Mss O129 S78 1914/15. Obstetrical applications.

Stanford University. School of Nursing.

Mss W181H C1S9. History.

Stanley, Leo Leonidas (1886–   ), resident physician, San Quentin Prison.

L626H E93. 236 v. (1931–1950). "Events of the day."

Mss L626H S77 1916–1923. 2 v. Correspondence.

Sullivan, Celestine.

Mss I982H S2S9. Manuscript material concerning San Francisco hospitals.

Sultan Ali tabib al Hurâsâni, Dustur ul-Iladji.

Mss Z268. "On remedies" (1526–1527).

Sweasey Powers, W. J.

Mss 14. 31 p. "Do the acute exanthemata represent different entities?"

Swift, Sidney Bourne.

Mss L154.9H H359 1891. Letter to Levi Cooper Lane, written from Molokai.

Tuhfat-ul-Mveminin der Tibb.

Mss Z265. 528 double p. (1669). Manuscript in Persian, "On medicine."

Ueber chronische Leyden im Allgemeinen.

Mss L48H M29. 459 l. Anonymous manuscript in German script.

University of the Pacific.

Mss H747H U58A2. Miscellaneous papers of the Medical Department, including records of students and instructors (1859–1883), account book, minutes, charter, and the resolution establishing the college.

Van Patten, Nathan

Mss U666H C53V2. Manuscript history, "Quinine: The first hundred years."

Warren, John Collins (1842–1927).

Mss 10. Letter (1910) to the chancellor of New York University, concerning William T. G. Morton, the discoverer of anaesthesia.

Wesson, Miley Barton (1881–?).

Mss H154H W51. 1 envelope. Miscellaneous papers; biographical material, especially regarding work as a consultant for the U.S. Army in Europe in 1949.

Whitney, James P. (1815–1880), president of the San Francisco Medical Society.

Mss H708H W61. Inaugural address.

Mss H117H.W61. Scrapbook containing his medical notes.

Wilbur, Ray Lyman (1875–1949), professor of physiology and medicine and dean of Cooper Medical College; university president, Stanford.

Mss 3. Correspondence and miscellaneous materials (1913–1954). Includes materials concerning medical education, the Baruch Committee on Physical Medicine, and the California Physicians' Service.

Wu, William.

Mss 18. 1 box, unprocessed. Papers.

Yusufi, Kitab-i-Tib.

Mss Z281. 166 l. (1782). Persian medical dictionary, titled "A book of medicine," prepared for Charles Grant.

ORAL HISTORIES

AHQP	Be.03	Crown Zellerbach	Be.05
AIP	Be.05	Cruess	Be.05
Alvarez	Be.05	Cutter Laboratories	Be.05
Alway	St.01	Dennes	Be.05
Amerine	Be.05	Dental History	Be.05
Appert	Be.05	Dental History	SF.02
Arnold	Be.05	DuBridge	Be.05
Arnstein	Be.05	Eddy Tree Breeding	Be.05
Beals	Be.05	Edmonston	Be.05
Berry	St.01	Evans, H. M.	Be.05
Berry, S.	Be.05	Farnsworth	St.01
Birge	Be.05	Fisher	Be.05
Blaney	Be.05	Foster	Be.05
Bloch	Be.05	Fowler	Be.05
Blochmans	Be.05	Fritz	Be.05
Bolt	Be.05	Fuller	Be.05
Bookman	Be.05	Gamow	Be.05
Boot	Be.05	Gerbode	Be.05
Born	Be.05	Gilfillan	Be.05
Bowen	Be.05	Gillespie	Be.05
Bowles	Be.05	Greenstein	Be.05
Bowman	Be.05	Grendon	Be.05
Bracelin	Be.05	Heintz	Be.05
Bradbury	Be.05	Helmholz	Be.05
Brobeck	Be.05	Henderson	Be.05
Bryant	Be.05	Hildebrand, J. H.	Be.05
Buttner	Be.05	Hinds	Be.05
Calvin	Be.05	Hodge	Be.05
Camp	Be.05	Hoover, William	St.01
Campbell, K.	Be.05	Jahns	St.01
Chamberlain	Be.05	Joslyn	Be.05
Chance	Be.05	Kaiser Permanente	Be.05
Chandler	St.01	Kamen	Be.05
Chaney	Be.05	Keen	St.01
Clark	Be.05	King	Be.05
Clark	St.01	Knudsen	Be.05
Coffman	Be.05	Kotcher	Be.05
Crandall	Be.05	Kroeber-Quinn	Be.05
		Krueger, M.	Be.05
		Kurti	Be.05
		Langelier	Be.05
		Leighton	Be.05
		Lennette	Be.05
		Lenzen	Be.05
		Lewis, R.	Be.05
		Libby	Be.05
		Livingood	Be.05
		Lofgren	Be.05
		Lowdermilk	Be.05
		Manley	Be.05
		Mark	Be.05

Mathias	Be.05	Wiggins	St.01
May	Be.05	Wilbur	St.01
Mayall	Be.05	Wilson	Be.05
McCulloch	Be.05	Winkler	Be.05
McGaughey	Be.05	Woodyard	Be.05
McLaughlin	Be.05	Wooldridge	Be.05
McMillan, E.	Be.05	Wyckoff	Be.05
Medical College	Be.05	York	Be.05
Metcalf	Be.05		
Meyer	Be.05		
Miller, L.	Be.05		
Miller, S.	Be.05		
Mirov	Be.05		
Moore	Be.05		
Mrak	Be.05		
Myers	Be.05		
Nelson	Be.05		
Olmo	Be.05		
Oppenheimer, F.	Be.05		
Packard	Be.05		
Perry	Be.05		
Poniatoff	Be.05		
Poppe	St.01		
Porter	Be.05		
Rawn	Be.05		
Rowntree	Be.05		
Schlesinger Library	Be.05		
Schreiber	Be.05		
Scofield	Be.05		
Scott, F.	Be.05		
Scott, K.	Be.05		
Seaborg	Be.05		
Senn	Be.05		
Shane	Be.05		
Shumate	Be.05		
Sinclair	Be.05		
Siri	Be.05		
Smith, C. S.	Be.05		
Spieth	Be.05		
Spinrad	Be.05		
Stanford University	St.01		
Stolz	St.01		
Storer	Be.05		
Tchelistcheff	Be.05		
Terman, F.	St.01		
Terman, F.	Be.05		
Thornton	Be.05		
Underhill	Be.05		
Van Dyke	Be.05		
Warren	Be.05		
Webster	Be.05		
Wiegand	Be.05		

INDEX

" | " denotes a repeated archive code.

Abrams, Albert, St.06 Alvarez

Abrams, Leroy, St.01 Abrams

Abu-Hamid Muhammed, St.06 Abu-Hamid Muhammed

Accademia del Cimento, Be.05 Accademia

Accelerators, Be.01; Be.02 Lawrence; Be.05 Brobeck | Goldhaber | Lawrence | Livingood | Maker | L. Marshall | Thornton; St.01 Hansen; St.02 Elwell; St.04

Acoustics, Be.05 Knudsen

Adams, Frank, Be.02 UCB, Agriculture; Be.05 Adams

Admiralty Library, London, Be.05 Demarcaciones | Drawings | Thompson | Ulloa

Aeronautics and astronautics, Be.05 Kotcher; Sa.02 Hiller; St.01 Siefert; St.02 Hopkins | Scientists and science

Agassiz, Louis, Be.05 Tuttle

Agency for International Development, St.06 Lee

Agnew, Harold Melvin, St.03 Agnew

Agriculture, Be.02 UCB, Agriculture; Be.05 Adams | Babcock | Blaney | Bonar | Burbank | Jaffa | P. Jordan | Lowdermilk | Packard | Wickson; Da.01 Audio-visual | Butterfield | Fairbanks | Ferry-Morse | Hunt | Scheuring | Voorhies | Wickson; Sa.02 Natomas Company; SC.03; St.03 Agriculture

———, and irrigation, Be.05 Mead | Tibbetts. *See also* Water resources

———, and technology. *See* Technology

Agronomy, Da.01 Madson | Peterson

Air ions, Be.05 A. Krueger Aitken, Robert, SC.01 Aitken

Akademiya Nauk S.S.S.R., St.01 Poppe

Al Ibrahimee, St.06 Al Ibrahimee

Al Kourashi, St.06 Al Kourashi

Albertus, Carl, Be.05 Albertus

Alchemy, Arabic, St.06 Moustafa ibn Halil us-Safiee

Alexander, Annie M., Be.02 UCB, Vertebrate Zoology; Be.05 Alexander

Allen, Joel A., Be.05 Allen

Alsos mission, St.03 Pash

Alvarez, Luis F., St.06 Alvarez

Alvarez, Luis W., Be.05 Alvarez | American Institute of Physics

Alvarez, Walter C., St.06 Alvarez

Alway, Robert, St.01 Alway

American Association for the Advancement of Science, Be.05 Treadwell; St.01 J. Branner

American Association of University Professors, St.01 G. Marx

American Beekeeping Federation, Da.01 American Beekeeping Federation

American Institute of Mining Engineers, St.02 Simplified spelling

American Institute of Physics, Be.05 American Institute of Physics | Kittel | Loeb | E. McMillan; SC.01

American Museum of Natural History, Be.05 A. Stone

American Ornithologists' Union, Be.05 Allen

American Philosophical Society, Be.05 American Philosophical Society

American Physical Society, Be.05 Birge; St.01 Bloch

American Psychological Association, St.01 Sears

American Relief Administration Association, St.03 Hutchinson

American Society of Mechanical Engineers, St.01 J. Branner

American Telephone and Telegraph, St.02 USA vs. AT&T. *See also* Bell Laboratories

Amerine, Maynard A., Be.05 Amerine; Da.01 Amerine | Vermouth Project

Ampex Corporation, Be.05 Poniatoff

Anatomy, St.06 Doremus | Hanzlik

Anderson, John A., Be.05 National Research Council

Animal husbandry, Da.01 Maillard

Animal science, Da.01 Richardson

Anthropology, Be.02 UCB, Anthropology; Be.05 Bascom | Beals | Gifford | Heizer | Kroeber | Kroeber-Quinn | Lowie | C. Merriam | Powers | Spier | U.S. Work Projects Administration

Antivivisection, Be.02 Anti-vivisection initiatives; Da.01 Richardson; St.01 R. Wilbur

Antoninus, Brother, Be.05 Everson

Apiculture, Da.01 American Beekeeping Federation | Apiculturalists | Apiculture subject files | Audio-visual | Beekeeping | California State Beekeepers | Eckert | Harbison | Laidlaw | McCubbin | Richter | Watkins. See also Entomology

Appert, Kurt E., Be.05 Appert

Aquariums, St.02 Seale

Archæological Institute of America, Be.05 Norton | Sparkman Archæology, Be.05 Bandelier | Cook | Deans | Fava | Heizer | Lenzen | C. Merriam | Norton | Schumacher | Sparkman | U.S. Work Projects Administration | Voy

Archive for History of Quantum Physics, Be.03; Be.05

Argonne National Laboratory, Be.05 Livingood; SF.02 Patt

Aristotle, Be.05 Levi ben Gershon

Arkansas, St.01 G. Branner

Arnold, James R., Be.05 Arnold

Arnot, Philip H., SF.02 Arnot

Arnstein, Lawrence, Be.05 Arnstein

Ashburner, William, Be.05 Ashburner

Ashcroft, Glenn B., Be.05 Ransom

Ashton, John, Be.05 Ashton

Astrology, Be.05 Everson

Astronautics, St.03 Beggs | Weakley. See also U.S. National Aeronautics and Space Administration

Astronomical Society of the Pacific, Be.05 Astronomical Society

Astronomy, Be.02 UCB, Astronomy | Regents; Be.05 Astronomical Society | Auhagen | W. Campbell | Cavaciocchi | Dieter | Drury | Einarsson | Galloway | Hale | Herschel | Hill | Leuschner | Minkowski | M. Sauer | Shane | Struve | Ulloa | Weitbrecht; SC.01; St.02 Johnson | Stanford University, Solar Observatory; St.02 Brasch | Miscelanea | See

Astrophysics, Be.05 American Institute of Physics | Hale | Minkowski

Atlantic cable, Be.05 Field

Atlases, Be.05 South Sea Waggoner. See also Cartography; Geography; Navigation

Audubon Association, Be.05 Audubon Association

Audy, J. Ralph, SF.02 Audy

Aughey, Samuel, Be.05 Aughey

Auhagen, Wilhelm, Be.05 Auhagen

Austin, Gladys, Be.05 Eddy Tree Breeding

Automobile design, Be.05 Doble

Autopsies, St.06 Ophüls

Autotherapy, St.06 Duncan

Averroes, Be.05 Levi ben Gershon

Aviation, Sa.02 Hiller; St.02 Hopkins. See also Aeronautics; Transportation

Babbage, Charles, Be.05 Babbage

Babcock, Ernest, Be.05 Babcock

Babcock, H. D., Be.05 National Research Council

Bacteriology, Be.05 A. Krueger | Meyer; SF.02 Harris

Bahr al Gauhar, St.06 Bahr al Gauhar

Bailey, Stanley F., Da.01 Bailey

Baja California, Be.05 Chickering. See also Mexico

Baldwin, John S., Be.05 Baldwin

Balloons, and scientific research, Be.05 Lee

Bancroft, Hubert Howe, Be.05 A. Goldschmidt | McCarthy

Bandelier, Adolphe F. A., Be.05 Bandelier | Norton

Banks, Harvey O., Be.05 Banks

Barkan, Adolf, St.06 Barkan

Barkan, Hans, St.06 Prentice

Barlow, Chester, Be.05 Barlow

Baruch Committee on Physical Medicine, St.06 R. L. Wilbur

Bascom, William R., Be.05 Bascom
Bates, Alfred, Be.05 LeConte
Bayard, Arnold A., Da.01 Bayard
Bélidor, Bernard Forest de, Be.05 Bélidor
Beals, Ralph, Be.05 Beals
Bears, Be.05 C. Merriam
Beattie, Margaret, Be.05 Beattie
Bechtel, W., S., and K., Be.05 Six Companies
Beekeeping. See Apiculture
Beggs, James, St.03 Beggs
Behaim, Martin, Be.05 Behaim
Bell Laboratories, Be.05 Buttner | Kittel | Knudsen; St.02 Kompfner; St.02 USA vs. AT&T; St.03 Gilman
Bennett, Merrill Kelly, St.01 Bennett
Bensley, Shad, St.06 Alvarez
Benson, Andrew Be.05 Benson
Berg, Paul, St.01 Berg
Berillon, Edgar, St.02 Berillon
Bernard, Claude, SF.02 E. Olmsted | J. Olmsted
Berne, Eric, SF.02 Berne
Bernstein, Benjamin, Be.05 Bernstein
Berry, Samuel Stillman, Be.05 S. Berry; St.01 Berry
Berthelot, Marcelin, St.02 Berillon
Bidwell, John, Be.05 Gray | Muir | Parry
Bikini Atoll, St.01 Whitaker; St.03 Lademan. See also Weapons testing
Billings' expedition, Be.05 M. Sauer
Billings, Bruce, St.03 Billings
Biochemistry, Be.05 Calvin | Fox | Hassid | Kamen | Koshland | Morgan | Stanley; Da.01 Guymon; St.01 Berg | Luck | Richardson; St.03 Luck
Biodynamics, chemical, Be.05 Calvin
Biology, Be.01; Be.05 H. M. Evans | Thomson; Be.06; St.01 Abrams | Campbell | Ehrlich | Stanford University, Biology | Whitaker | Wiggins; St.02 Seale. See also fields within biology

———, marine, Be.05 S. Berry | Fox | Hedgpeth | Ritter; St.01 Berry | Jenkins | McFarland | A. Moore | G. Moore | Stanford University, Biology; St.02 Ricketts
———, molecular, Be.02 Fraenkel-Conrat; Be.05 Stanley
Biophysics, Be.05 Anger | Grendon | Myers | Siri | Strait | R. Williams. See also Physics, medical; Medicine, nuclear
Biostatistics, Be.05 Yerushalmy
Birge, Raymond, Be.05 Birge
Birmingham, University of, Be.05 Boot; St.01 Kompfner
Birth control, Be.05 Schlesinger Library; St.06 Pamphlets
Bishop Museum, Honolulu, Be.05 Usinger; St.02 Seale
Bishop, Louis, Be.05 Bishop
Black, Joseph, St.06 Black
Blackwelder, Richard, St.03 Blackwelder
Blaney, Harry, Be.05 Blaney
Bloch, Felix, Be.05 American Institute of Physics; St.01 Bloch | Stanford University. News and Publications
Blochman, Ida, Be.05 Blochmans
Bodonus, Carlus Antonius, St.06 Bodonus
Boethius, Be.05 Testi
Bohart, Richard, Da.01 Bohart
Bohr, Niels, Be.03; Be.05 Bohr
Bok, Bart J., Be.05 Dieter
Bolander, Henry N., Be.05 Lesquereux
Bolt, Richard, Be.05 Bolt
Boltzmann, Ludwig, Be.05 Jaffé
Bonar, Lee, Be.05 Bonar
Boodt, William A., Be.05 Boodt
Bookman, Max, Be.05 Bookman
Boot, Henry A. H., Be.05 Boot
Born, James L., Be.05 Born
Born, Leonard L., Da.01 Born
Borosini, Dr., St.06 Fletcher
Bošković, Rudjer Josip (Boscovich), Be.05 Bošković

168

Botanical Society, St.01 Stanford University, Botanical Society

Botany, Be.05 Blochmans | Bonar | Bracelin | Brewer | Burbank | Burke | Chaney | Chickering | Cremer | Franceschi | Goodspeed | Gray | Hall | M. Hildebrand | Howard | James | Jepson | Judah | Lesquereux | Mathias | Metcalf | Mexía | Muir | Rowntree | Schneider | F. Scott | Setchell | Treadwell | Yoakum; Da.01 Botanical prints | Crafts | Esau | Ferry-Morse | Index seminum | Richter | Stebbins; Sa.01; St.01 Abrams | Campbell | D. Jordan | Peirce | Stanford University, Botanical Society | Wiggins; St.06 Bodonus. *See also* Horticulture

Botta, Paul, Be.05 Botta

Botulism, St.06 Dickson | Geiger

Boulder Dam, Be.05 Six Companies

Bourguin, Burnice, St.01 Bourguin

Bowen, Ira Sprague, Be.05 American Institute of Physics

Bowles, Edward, Be.05 Bowles

Bowman, Karl M., Be.05 Bowman

Bracelin, Nina Floy, Be.05 Bracelin | Mexiia

Bradbury, Norris E., Be.05 Los Alamos

Bragg, W. H. and W. L., Be.03

Branner, George, St.01 G. Branner

Branner, John, St.01 J. Branner

Brasch, Frederick E., St.02 Brasch

Bray, William C., Be.05 Bray

Brazil, Be.05 Drawings

Breit, Alfons, St.06 Breit

Brera Observatory, Be.05 Bošković

Brewer, William H., Be.05 Ashburner | Brewer | California Geological Survey | Hoffmann

Bridges, Lyman, Be.05 Bridges

Bridges, St.02 Strauss. *See also* Engineering

Brigham, Charles, SF.02 Brigham

Brisbane, Arthur, Be.05 Byerly

British Association for the Advancement of Science, Be.05 Lee

British Library, Be.05 South Sea Waggoner

Brobeck, Wiliam M., Be.05 Brobeck

Brode, Bernice Bidwell, Be.05 Brode

Brooks, Frederick A., Da.01 Brooks

Brown, Adelaide, St.06 A. Brown

Brown, Benjamin, St.06 B. Brown

Brown, John Carter, Library, Be.05 South Sea Waggoner

Brown, Philip, St.06 P. Brown

Bryant, Harold, Be.05 Bryant

Bubble chamber, Be.05 Alvarez

Buenos Aires, Be.05 Proyecto

Burbank, Luther, Be.05 Burbank | Howard | Welch | Wickson; Da.01 Wickson

Burbridge, Thomas N., SF.02 Burbridge

Burke, Joseph, Be.05 Burke

Burthmann, E., St.02 Ritter

Butler, Charles, St.03 Butler

Butterfield, Harry, Da.01 Butterfield

Butters, Charles, Be.05 Butters

Buttner, Harold, Be.05 Buttner

Byerly, Perry, Be.05 Byerly | Seismological Society

CERN, St.01 Bloch

COMSAT (Communications Satellite Corporation), Be.05 COMSAT

Čebiš, František Rudolf, Da.01 Čebiš

Cabrillo's grave, Be.05 Heizer

Cajori, Florian, Be.05 Cajori

California, State of, Sa.01

———, Archives, Sa.01

———, Engineer, Sa.02 Hall

———, Food and Drug Laboratory, Be.04; Be.05 Jaffa

———, Library, Sa.02

———, Department of Public Health, Be.05 Warren

———, Bureau of Sanitary Engineering, Be.05 Gillespie

California Academy of Sciences, Be.05 Davidson | Dexter | Martens | Ransom |

Rowntree | Sparkman; SF.01; St.01 J. Branner; St.02 Seale
California Council for the Blind, Be.05 Perry
California Geological Survey, Be.05 Ashburner | Brewer | California Geological Survey | Gilman | Hoffman | J. L. LeConte | Yates
California Institute of Technology, Be.05 Hale | National Research Council
California Packing Corporation, St.06 Dickson
California Petroleum Company, St.02 Peckham
California Physicians' Service, St.06 R. L. Wilbur
California Public Health, Be.05 California Public Health
California Society for the Promotion of Medical Research, SF.02 California Society; St.06 California Society
California State Beekeepers Association, Da.01 California State Beekeepers
California State Telegraph Company, Sa.02 Moss
California Water Project, Be.05 Banks
Calvin, Melvin, Be.05 Calvin
Cameron, Donald R., Be.05 Cameron
Camp, Charles L., Be.05 Camp
Campbell, Douglas, St.01 Campbell
Campbell, Kenneth, Be.05 Campbell
Campbell, William Wallace, Be.02; Be.05 Campbell; SC.01 Campbell | Keeler | Leuschner
Canada, Be.05 Baldwin | Burke | Deans
———, Geological Survey, Be.05 Canada Geological Survey | A. Lawson
Candolle, A. de, Be.05 Lombard
Canners' League, St.06 Dickson
Caplan, Herb, Da.01 Caplan
Carbon-14, Be.05 Calvin | Kamen | Libby
Carl-Ferdinand Universität, Prague, Be.05 Einstein
Carnegie Institute of Technology, Be.05 Stern

Carpender, Jack, Be.05 Eddy Tree Breeding
Carson Hill Gold Mining Corporation, Sa.01 Carson Hill
Cartography, Be.05 Behaim | Drawings | A. Goldschmidt | South Sea Waggoner; Da.01 Power/Drake research; SC.03; St.06 Munshi-Mazhar-al-Din. *See also* Geography; Surveys
Castaño, Arturo, Be.05 Proyecto
Caterpillar Tractor Company, Da.01 Higgins
Catholic University of America, Be.05 Henderson
Cavaciocchi, Michele, Be.05 Cavaciocchi
Celsian Association, St.06 Celsian Association
Census, Sa.01
Cerda, Tomás, Be.05 Cerda
Ceylon, St.06 Old Sinhalese Herbalist | Sinhalese Ola
Chadbourne, H. L., Be.05 Chadbourne
Chamberlain, Owen, Be.05 Chamberlain
Chambers, Willie Lee, Be.05 Chambers
Chance, Ruth, Be.05 Chance
Chandler, Loren, St.01 Chandler; St.06 Chandler
Chaney, Ralph, Be.05 Chaney
Chappe, l'abbé, Be.05 Pauly
Chemical industry, St.03 Interessengemeinschaft Farbenindustrie Aktiengesellschaft
Chemistry, Be.01; Be.02 UCB, Chemistry | Hildebrand; Be.05 Braceline | Bray | Calvin | Crandall | E. Fischer | Giauque | Hassid | J. Hildebrand | Huffman | Jaffa | K. Jenkins | Langmuir | G. N. Lewis | Rising; Da.01 Nelson; St.01 Luck | Swain; St.02 Berillon | Notebook; St.03 Luck; St.06 Black | Bodonus | B. Brown
———, industrial, Be.05 Fischer | S. Tibbetts; Da.01 Caplan; St.03 Interessengemeinschaft
———, nuclear, Be.01; Be.05 Arnold | Kamen | G. N. Lewis | Libby | S. Miller | Schulz | Seaborg
———, pharmaceutical, Be.05 Cutter Laboratories; SF.02 Riegelman | Strait

———, physical, Be.05 Giauque | J. Hildebrand | Kurti | G. N. Lewis | Villars; St.01 Findlay

Chevalier, François, Be.05 Chevalier

Chicago, University of, Be.05 C. S. Smith

Chickering, Allen Lawrence, Be.05 Chickering

Child guidance, Be.05 Senn

Childs, Herbert, Be.02 Childs

China, Be.05 Louderback; St.03 Gunn

Chirac, Pierre, St.02 Notebook

Chodorow, Marvin, Be.05 Chodorow

Christy, Samuel, Be.05 Christy

Churchill, Alexander, Be.05 Churchill

Churchman, C. West, Be.05 Churchman

Clarendon Laboratory, St.01 Kompfner

Clark, Esther Bridgman, St.01 Clark

Clark, Nathan C., Be.05 Clark

Clarke, William J., Be.05 Chadbourne

Clement, Lewis M., Be.05 Clement

Cluff, Benjamin, Be.05 Roberts

Coast Manufacturing and Supply Company, Be.05 Coast Manufacturing

Coffman, John D., Be.05 Coffman

College of Rome, Be.05 Bošković

Colorado Committee for Environmental Information, St.03 Metzger

Columbia University, Be.05 James

Commandino, Federigo, Be.05 Testi

Communications satellites, Be.05 COMSAT

Computers and computer science, Be.05 Babbage | Crandall | Taub; St.01 ACME Project | Forsythe | Knuth

Conchology, Be.05 Gulick | Sowerby | Yates; St.01 E. Jordan; St.02 Lewis | Oldroyd

Confederate States of America, Be.05 Yoakum

Conservation, Be.05 Bryant | Chaney | Fritz | J. Hildebrand | Lowdermilk | R. Marshall | McCulloch | J. Merriam | Muir | Rawn | Sierra Club | Siri. *See also* Sierra Club

Constance, Lincoln, Be.05 Bonar | Jepson

Conti, Giovan Stefano, Be.05 Bošković

Contraception. *See* Birth control

Cook, Sherburne Friend, Be.05 Cook | Heizer

Cooksey, Donald, Be.01; Be.05 Lawrence

Cooper Medical College, St.01 R. L. Wilbur | Stanford University, President; St.06 Alvarez | Cooper Medical College | Cox | Higgins | Lane | Prentice | R. L. Wilbur. *See also* Stanford University

Cooper Ornithological Club/Society, Be.05 Bishop | Chambers | Cooper Ornithologica Society | Grinnell | A. Miller | Tyler

Cooper, James, Be.05 J. Cooper

Coordinated Investigations of Micronesian Anthropology (CIMA), SC.03

Cornell University, Da.01 Decker

Corti, Egon Caesar, Da.01 Corti

Cosmography, Be.05 Thevet

Cotes, Roger, Be.05 Cotes

Cove/Bakken Advertising Company, Da.01 Cove/Bakken

Cox, Alvin J., St.06 Cox

Crafts, Alden, Da.01 Crafts

Craig, Thornton, Da.01 Craig

Cranberg, Lawrence, St.03 Cranberg

Crandall, Howard, Be.05 Crandall

Crawford, Albert, St.06 Crawford

Creation Research Society, Be.05 Lammerts

Cremer, John, Be.05 Cremer

Crown Zellerbach, Be.05 Crown Zellerbach

Cruess, William, Be.05 Cruess; Da.01 Cruess

Cullen, Stuart, SF.02 Cullen

Cumming, William G., Be.05 Eddy Tree Breeding

Cushing, Clinton, St.06 Cushing

Cutter Laboratories, Be.05 Cutter Laboratories

Cutter, Lawrence, St.01 Cutter

Cyclotrons, Be.01; Be.02 UCB, Radiation Laboratory; Be.05 Brobeck | Lawrence |

Livingood | Woodyard. *See also* Accelerators; Lawrence, Ernest O.; UCB, Lawrence Berkeley Laboratory

D'Alembert, Jean LeRond, St.02 Norman

Dakin, Frederick, Be.05 Dakin

Dall, William H., St.02 Oldroyd

Dana, Samuel T., Be.05 Dana,

Daniel, Book of, Be.05 Newton

Darling, William, St.03 Darling

Darwin, Charles, Be.05 Darwin

Davidson, George, Be.05 Davidson | J. Lawson

Dawson, Ernest, Be.05 Jepson

De Forest, Clara, St.06 De Forest

De Forest, Lee, Be.05 Ashton | H. Stone | Simon

Deans, James, Be.05 Deans

Decker, Frank, Da.01 Decker

Dennes, William R., Be.05 Dennes, Dentistry, Be.05 Dental History; SF.02 Dental History | Fleming | Jungck

Depression, the Great, Be.02 U.S. Work Projects Administration

Derleth, Charles, Be.05 Derleth

Dermatology, St.06 Cooper Medical College | Hebra

Deutsche Forschungsgemeinschaft, St.03 Deutsche Forschungsgemeinschaft

Dexter, Henry, Be.05 Dexter

Dickson, Ernest, St.06 Dickson

Dieter, Nannielou, Be.05 Dieter

Dixon, Robert E., SF.02 Dixon

Doble, Abner, Be.05 Doble

Doble, Robert, Be.05 Doble

Dodd, Coleman, Be.05 Dodd

Dolger, David, St.06 Dolger

Dollar (Robert) Company, Be.05 Dollar | Heintz

Doremus, R. Ogden, St.06 Doremus

Drake, Sir Francis, Da.01 Power/Drake research

Drinker, Cecil, St.06 Drinker

Drug abuse, Be.05 H. Jones

Drury, Aubrey, Be.05 Drury

DuBridge, Lee Alvin, Be.05 American Institute of Physics

Duke-Elder, William Stewart, St.06 Duke-Elder

Duncan, Charles H., St.06 Duncan

Dunphy, John E., SF.02 Dunphy

Durbin, Patricia W., Be.05 Durbin

Earth sciences, Be.01; St.01 Jahns | Park. *See also* Seismology

Earthquakes, Be.05 Derleth | Stratton; St.01 Earthquake collection | Stanford University, Commission of Engineers; St.06 Gould

Eastwood, Alice, SF.01

Eastwood, John, Be.05 Eastwood

Eckert, John, Da.01 Eckert

Ecology, Be.05 Tchelistcheff; Sa.01; SF.02 Audy; St.01 Ehrlich. *See also* Conservation

Economics, Be.06; Da.01 Voorhies; SC.03

Eddy Tree Breeding Station, Be.05 Eddy Tree Breeding

Edinburgh, University of, St.01 Findlay

Edison, Thomas Alva, Da.01 Edison

Edmonston, Robert M., Be.05 Edmonston

Education, agricultural, Da.01 Hutchison | Ryerson | Shields | University Archives

Education, medical. *See* Medical education

Education, science, Be.05 J. Hildebrand | Purcell; in Paris, Be.05 Paris

Ehrenfest, Paul, Be.03

Ehrlich, Anne H., St.01 Ehrlich

Ehrlich, Paul R., St.01 Ehrlich

Einarsson, Sturla, Be.05 Einarsson

Einstein, Albert, Be.05 Einstein | Lenzen; St.02 Brasch

Einstein, Mileva, Be.05 Einstein

Einstein Project, St.01 Stanford University, News and Publications

Eisen, Gustav, SF.01

Eitel-McCullough, Be.05 Eitel-McCullough | McCullough

El Salvador, Be.05 M. Hildebrand

Electricity, St.02 Franklin

Electronics, Be.05 Albertus | Appert | Boot | Bowles | Buttner | Chadbourne | Chodorow | Clark | Dodd | Eitel-McCullough | Fuller | Heintz | Litton | L. Marshall | Moore | Perham | Silver | Terman | Varian Associates | Vasseur | Whinnery; St.01 F. Terman | Hansen | Rambo | Shockley | Stanford University, Electronics Laboratory | Webster; St.02 Elwell | USA vs. AT&T; St.03 Packard

———, and medicine, Be.05 Marg | Marmont | Weitbrecht

Elliott, George, Be.05 Elliott

Elliott, Harold F., Be.05 Perham,

Elliott, John Elbert, St.03 Elliott

Ellis, N. Randall, Be.05 Ellis

Eloesser, Leo, St.06 Eloesser

Elwell, Cyril F., St.02 Elwell

Emerson, William, Be.05 W. Emerson

Emerson, George H., St.03 Emerson

Energy, Be.01; Be.05 Legge. *See also* Power

Engels, Friedrich, St.03 Cranberg

Engineering, Be.02 UCB, Architects and Engineers | UCB, Engineering | UCB, Engineering, Science, and Management War Training Program; Be.05 Adams | Bélidor | Blaney | Brobeck | Churchill | Doble | Ellis | F. Tibbetts | Foster | Gaytes | Marmont | Pinger | Proyecto | Putnam | Randall | Reynolds | Rhodin | H. Stone; Be.06; Da.01 Holland Land Company; Sa.02 Hall | McCartney; St.01 Stanford University, Commission of Engineers | Stanford University, Engineering; St.02 Heintz | Hyde | Simplified Spelling Collection | Strauss | Sutro; St.03 Gilman | Hoover | Poncelet | Rabbitt | Stuart

———, agricultural, Be.05 Mead | Packard; Da.01 Brooks | Jacobsen | Miller-Lux | Neubauer | Walker

———, civil, Be.05 Banks | Bookman | Bridges | Derleth | Eastwood | Edmonston | Hinds | Langelier | Manson | Mitchell; Sa.02 Givan; St.01 C. Marx | Morris; St.03 Gorton | Stevens

———, construction, Be.05 Six Companies

———, electrical, Be.05 Appert | Clark | Fuller | L. Marshall | Pollard | Powell | Silver | Terman | Whinnery | Woodyard; St.01 ———, electrical, St.01 Bourguin | G. Branner | Gorman | W. Hoover | Rambo | Ryan | Shockley | Stanford University, Ryan High Voltage Laboratory | F. Terman

———, hydraulic, Be.05 McCarthy

———, mechanical, Be.05 Rix | J. Smith; St.01 Cutter | Fuchs | Gorman | G. Marx | Jacobsen | Skilling | Stanford University, Mechanical Engineering

———, mining, Be.02 UCB, Mining Association; Be.05 Butters | Dakin | Hohl | McLaughlin | Morris | Sizer; Sa.02 Sutro; St.02 Hurst | Sutro

———, petroleum, Be.05 Maker | Morris; St.02 Peckham

———, radio, Be.05 Clement | Fisher | H. Miller | Scofield | Simon | Sinclair

———, sanitary, Be.05 Gillespie | Langelier | McGaughey | Rawn; Sa.01

———, in the Soviet Union, St.03 Darling | Emerson | Farquhar | Foss | Gorton | Hoskin | Hutchinson | Johnson | Russian subject collection | Starr | Stepanov | Stevens | Stuart | U.S. Advisory Commission | Witkin

ENIAC, Be.05 Taub

Entomology, Be.05 Essig | Popenoe | Spieth | Usinger | Woodworth; Da.01 Bailey | Bohart | Eckert | Freeborn | Laidlaw; Sa.01. *See also* Apiculture

Environmental studies, Be.01; St.03 Metzger. *See also* Ecology

Epidemiology, Be.05 Beattie | G. Jones | Mexico, Archivo General | Richter; SF.02 Audy | Hooper Foundation; Sa.01; St.06 Geiger | Pont. *See also* Public health

Epstein, Paul, Be.05 Wise

173

Esau, Katherine, Da.01 Esau

Essig, Edward, Be.05 Essig

Ethnology, Be.02 UC, Anthropology; Be.05 Botta | Hodge | Pinart. *See also* Anthropology; Linguistics

Euclid, Be.05 Euclides

Eugenics, Be.05 S. Holmes; St.01 D. Jordan | L. Terman. *See also* Genetics

Euthanasia, Da.01 Richardson

Evans, Griffith C., Be.05 G. Evans

Evans, Henry, Da.01 Botanical prints

Evans, Herbert McLean, Be.02 UCB, History of Science; Be.05 H. Evans

Everman, B. W., SF.01

Everson, William (Brother Antoninus), Be.05 Everson

Evolution, Be.05 Darwin | Gulick | S. Holmes | Spieth; St.01 Whitaker

Exactus Photo-Film Corporation, St.02 Exactus

Expeditions, Be.05 Farquhar | James | Thevet | Thompson; Da.01 Power/Drake

——, astronomical, Be.05 Pauly | Ulloa

——, scientific, Be.05 Botta | Burke | M. Hildebrand | Kofoid | C. Merriam | Palmer | Roberts | M. Sauer. *See also* Surveys

Exploratorium, Be.05 Exploratorium

Faber, Harold Kniest, St.06 Faber

Faber, Mary, St.06 Faber

Fairbanks, William, St.01 Stanford University, Interviews

Fairbanks, James P., Da.01 Fairbanks

Farnsworth, Paul, St.01 Farnsworth

Farnsworth, Philo T., Be.05 Lippincott,

Farquhar, Francis P., Be.05 Brewer | Farquhar

Farquhar, Percival, St.03 Farquhar

Fauche-Borel, Louis, St.02 Fauche-Borel

Fava, Florence M., Be.05 Fava

Federal Telegraph Company, Be.05 Fisher | Fuller | H. Miller | Pratt

Fermi, Enrico, Be.05 Segrè

Fermi, Laura, Be.05 Brode

Ferry-Morse Seed Company, Da.01 Ferry-Morse

Fidler, Harold, Be.05 Fidler

Field, Cyrus, Be.05 Field

Findlay, Alexander, St.01 Findlay

Fire, control of, Be.05 Coffman; prevention, Be.05 Metcalf

Fischer, Emil, Be.05 E. Fischer

Fischer, Hermann, Be.05 E. Fischer

Fischer, Otto, Be.05 O. Fischer

Fisher, Gerhard, Be.05 Fisher

Fisher Research Laboratory, Be.05 Fisher

Fisheries, St.02 Seale

Fleming, Willard C., SF.02 Fleming

Fletcher, Horace, St.06 Fletcher

Floyd, Richard S., SC.01 Floyd | Fraser | Holden

Folklore, Be.05 Bascom. *See also* Ethnology

Food science, Be.05 Bracelin | Joslyn | Mrak; Da.01 Born | Cruess | Mrak | Steinmetz

Food technology, Be.05 Cruess | Mrak; Da.01 Pomology; Sa.01; Sa.02 Hiller; St.06 Dickson. *See also* Botulism; Nutrition

Foothill Electronics Museum, ii; Be.05 Perham

Forestry, Be.05 Coffman | Crown Zellerbach | Dana | Eddy Tree Breeding | Fritz | M. Krueger | Leopold | Lowdermilk | McCulloch | Metcalf | Mirov | Nelson | Vaux; Da.01 Marsh; St.01 Bracewell

Forsythe, Alexandra, St.01 Forsythe

Forsythe, George, St.01 Forsythe

Foss, F. F., St.03 Foss

Foster, Herbert, Be.05 Foster

Fowler, William, Be.05 American Institute of Physics

Fox, Denis, Be.05 Fox

Fraenkel-Conrat, Heinz Ludwig, Be.02 Fraenkel-Conrat

Fraiberg, Selma H., SF.02 Fraiberg
France, king of, Be.05 Pauly | Thevet
Franceschi, Francesco, Be.05 Franceschi
Franklin, Benjamin, St.02 Franklin
Fraser, Thomas E., SC.01 Floyd | Fraser | Holden
Freeborn, Stanley, Da.01 Freeborn
Fritz, Emanuel, Be.05 Fritz
Froman, Darol. See Be.05 Los Alamos
Fuchs, Henry O., St.01 Fuchs
Fuller, Leonard, Be.05 Fuller. See also Federal Telegraph Company
Fuller, George, Be.02 Fuller

Galilei, Galileo, Be.05 Galileo
Galloway, John Debo, Be.05 Galloway
Gamow, George, Be.05 American Institute of Physics
Gardner, Max, Be.05 Gardner
Gayet-ul-Beyan fil Tib, St.06 Gayet-ul-Beyan fil Tib
Gaytes, Herbert, Be.05 Gaytes
Geiger, Jacob, St.06 Geiger
General Electric Research Laboratory, Be.05 Langmuir
General Radio Company, Be.05 Sinclair
Genetics, Be.05 American Philosophical Society | Babcock | Cameron | R. Goldschmidt | Stebbins; St.01 Shockley
———, plant and forest, Be.05 Eddy Tree Breeding | J. Jenkins | Olmo
Geneva, University of, Be.05 Lombard
Geography, Be.05 Behaim | Cavaciocchi | Davidson | Mirov | Pauly | Pinart | C. Sauer | M. Sauer | Spink; Da.01 Power/Drake; St.02 Pacific Geographic Society; St.06 Munshi-Mazhar-al-Din | Shal mardan al Mustawf
Geological Society of American Universities, St.01 Geological Society
Geology, Be.02 Turner | Ashburner | Aughey | Brewer | California Geological Survey | Canada Geological Survey | Cooper | Gilman | Hodge | Hoffmann | James | Kimble | Lawson | LeConte |

Louderback | Matthes | McLaughlin | Muir | Palache | Seismological Society | H. Stone | S. Tibbetts | Turner; Da.01 Gilmore | Skaar Mining; Sa.01; St.01 Geological Club | Geological Society | Branner | Park | Stanford University, Earth Sciences, Geology; St.02 Peckham | Scientists and science; St.03 Elliott
Geometry, Be.05 Euclides | Pardies | Rabuel | Regnault
Geophysics, Be.05 Turner
George, Thomas C., Be.05 George
Gerbode, Frank, Be.05 Gerbode
Ghigath-al-Din, St.06 Ghigath-al-Din
Giauque, William, Be.05 Giauque | Kurti
Gibbons, Henry, St.06 Gibbons | Higgins
Giessen, University of, Be.05 Jaffé
Gifford, Edward, Be.05 Gifford
Gilbert, Charles, St.01 Gilbert
Gilfillan, S. Colum, Be.05 Gilfillan
Gilinsky, Victor, St.03 Gilinsky
Gillespie, Chester G., Be.05 Gillespie
Gilman, Daniel, Be.05 Gilman
Gilman, G. W, St.03 Gilman
Gilmore, Arthur E., Da.01 Gilmore
Gimbel lectures, Be.05 Kroeber
Givan, Albert, Sa.01 Givan
Globe Wireless Company, Be.05 Dollar collection
Göttingen, University of, Be.05 Bohr
Gofman, John, Be.05 Gofman
Golden Gate Bridge, Be.05 Derleth; St.02 Strauss
Goldhaber, Sulamith, Be.05 Goldhaber
Goldschmidt, Albert, Be.05 A. Goldschmidt
Goldschmidt, Richard, Be.05 R. Goldschmidt
Goodfellow, Hugh, St.06 Stanford University, Medical Center
Goodspeed, Thomas Harper, Be.05 Goodspeed
Gorman, Mel, Be.05 Bray
Gorman, William H., St.01 Gorman

175

Gorton, Willard L., St.03 Gorton
Gould, Agnes, St.06 Gould
Graham, Beardsley, Be.05 COMSAT
Grant, Charles, St.06 Yusufi
Gray, Asa, Be.05 Darwin | Gray | Tuttle
Grayson, Andrew Jackson, Be.05 Grayson
Greenstein, Jesse, Be.05 American Institute of Physics
Gregory, John, Be.05 Gregory
Grendon, Alexander, Be.05 Grendon
Griffith, Benjamin G., Be.05 Griffith
Grinnell Naturalists, Be.05 Grinnell Naturalists
Grinnell, Joseph, Be.05 Alexander | Barlow | Grinnell | Tyler
Grinnell, Hilda, Be.05 Grinnell
Guerlac, Henry, Be.05 Guerlac
Gulick, John T., Be.05 Gulick
Gunn, Herbert, St.06 Gunn
Gunn, Selskar M., St.03 Gunn
Guymon, James Fuqua, Da.01 Guymon
Gynecology, St.06 Cooper Medical College | Cushing
HEP, St.04 High-energy physics
Hack[e], William, Be.05 South Sea Waggoner
Hale, George Ellery, Be.05 Hale
Hall, Harvey, Be.05 Hall
Hall, William, Sa.01 Hall
Halls of Aesculapius, St.06 Halls of Aesculapius
Hamburg, University of, Be.05 Stern
Hamilton, Joseph G., Be.05 Durbin | Hamilton | Lawrence
Hansen, William, Be.05 Hansen; St.01 Hansen
Hanzlik, Paul, St.06 Hanzlik
Harbison, John Stewart, Da.01 Harbison
Harris, Henry, SF.02 Harris
Hart, George H., Da.01 Hart

Hart, Henry H., Be.04
Harvard University, Be.05 Farquhar | Palache | Tuttle; St.01 Webster; St.02 Seale
———, Radio Research Laboratory, Be.05 Livingood | Terman | Sinclair; St.01 F. Terman
Harvey, A. McGehee, St.06 Rytand
Hassid, Zev, Be.05 Hassid
Health planning, SF.02 West Bay. *See also* Public health
Hearing impairment, electronic devices and, Be.05 Weitbrecht
Hebra, Ferdinand von, St.06 Hebra
Hedgpeth, Joel, Be.05 Hedgpeth
Heijenoort, Jean van, St.01 Kreisel
Heintz, Ralph, Be.05 Heintz | Scofield; St.02 Heintz
Heintz and Kaufman, Be.05 Heintz | McCullough
Heisenberg, Werner, Be.03
Heizer, Robert F., Be.05 Cook | Heizer | Sturtevant
Helicopter, Sa.02 Hiller
Helmholz, A. Carl, Be.05 Helmholz
Hendel, Charles, Sa.01 Hendel
Henderson, Malcolm, Be.05 Henderson
Herbals, St.06 Bahr al Gauhar | Bodonus | Old Sinhalese herbalist | Sinhalese herbal
Herbaria, Be.05 Jepson | Judah
Herschel, Sir John Frederick William, Be.05 Herschel
Hetch-Hetchy, Be.05 R. Marshall
Hewlett-Packard Company, Be.05 Terman; St.01 F. Terman; St.02 Elwell; St.03 Packard
Heynes, Henry, Be.05 Norton
Higgins, Alice, St.06 Higgins
Higgins, Floyd, Da.01 Higgins
Higgins, William M., St.06 Higgins
Hildebrand, Joel Henry, Be.02 Hildebrand; Be.05 J. Hildebrand

Hildebrand, Milton, Be.05 M. Hildebrand

Hilgard, Eugene, Be.05 Hilgard | Slate

Hill, Charles Barton, Be.05 Hill

Hiller, Stanley, Jr. and Sr., Sa.01 Hiller; Sa.02 Hiller

Hinds, Julian, Be.05 Hinds

Hiroshima-Nagasaki Publishing Committee, St.03 Hiroshima-Nagasaki

Histology, St.01 McFarland

Hitch, Charles, Be.02 Hitch

Hodge, Frederick, Be.05 Hodge

Hoffmann, Charles F., Be.05 Hoffmann

Hohl, L. J., Be.05 Hohl

Holden, Edward, SC.01 Floyd | Fraser | Holden | Keeler

Holland Land Company, Da.01 Holland Land

Holmes, Jack, Be.05 J. Holmes

Holmes, Samuel, Be.05 S. Holmes

Home economics, Be.05 Morgan. *See also* Nutrition

Homestake Mining Company, Be.05 McLaughlin

Honigbaum, Max, Be.04

Hooker, Sir William Jackson, Be.05 Burke

Hooper Foundation, SF.02 Audy

Hooper, George Williams, SF.02 Hooper | Hooper Foundation

Hoover, Herbert, Be.05 Seismological Society; St.01 R. Wilbur

Hoover, Theodore, St.03 Hoover

Hoover, William, St.01 W. Hoover

Hopkins Marine Station, St.01 Jenkins | McFarland | A. Moore | G. Moore | Hopkins Marine Station | Stanford University, Biology Department. *See also* Stanford University

Hopkins, Timothy, St.02 Hopkins

Horn, George, Be.05 Popenoe

Horticulture, Be.05 Burbank | Rowntree | Welch; Da.01 Butterfield | Ferry-Morse | Index seminum | Nursery and seed catalogs | Ryerson | Wickson. *See also* Botany; Agriculture

Hoskin, Harry L., St.03 Hoskin

Hospitals, Be.04; SF.02; St.06. *See also* Medicine, clinical

Howard, Walter, Be.05 Howard

Howard, Nelson J., St.06 Gunn

Huffman, Eugene, Be.05 Huffman

Humboldt, Alexander von, Be.05 Behaim

Hunt, Thomas, Da.01 Hunt

Hunter, William, St.06 Hunter

Huntington Library, Be.05 South Sea Waggoner

Huntington, Thomas W., St.06 Huntington

Hurst, George R., St.02 Hurst

Hutchinson, Lincoln, St.03 Hutchinson

Hutchison, Claude, Da.01 Hutchison

Hyde, William, St.02 Hyde

Hydraulics, Be.06; B.05 Cerda

Hydrography, Sa.02 Hall

Hydrology, Be.05 Lowdermilk; Be.06

Hydrostatics, Be.05 Cerda

Hypnosis, St.01 Stanford University, Hypnosis Laboratory; St.02 Berillon

Ichthyology, Be.05 D. S. Jordan | Needham; St.01 Gilbert | D. S. Jordan | Vanderbilt Foundation; St.02 Seale

Illustrations, botanical, Da.01 Botanical prints

Indiana University, St.01 J. Branner

Indians, Be.05 Kroeber | Kroeber-Quinn | Lowie | C. Merriam | Pinart | Powers | Spier | Sturtevant | U.S. Work Projects Administration; land claims, Be.05 Heizer | Kroeber

Industry, in the Soviet Union, St.03 Foss

Inman, Verne, SF.02 Inman

Instruments, scientific, Be.02 Fuller | Thacher; Be.05 Marg | Slate; SC.01; for surveying, Be.05 Martens

Intelligence testing, St.01 L. Terman | Shockley

Inter-Allied Technical Board, St.03 Johnson

Interessengemeinschaft Farbenindustrie Aktiengesellschaft, St.03 Interessengemeinschaft

Intergovernmental Oceanographic Commission, St.03 Intergovernmental Oceanographic Commission

International Astronomical Union, Be.05 Struve

International Business Machines Corporation (IBM), Be.05 Smathers

International Commission on Eugenics, St.01 L. Terman

International Telephone and Telegraph Corporation, Be.05 Buttner

International Union of Pure and Applied Physics: Be.05 International Union

Inventions, and inventors, Be.02 UCB, Patent Office; Be.05 Heintz | Muir | Rix | Smathers | Weitbrecht; Da.01 Edison; Sa.02 Hiller; St.02 Heintz; St.03 Kantor. *See also* Patents

Irrigation, Be.02 UCB, Agriculture; Be.05 Doble; Da.01. *See also* Agriculture

Italy, science in, Be.05 Accademia del Cimento | Galileo; St.02 Miscelanea

Jack and Heintz, Inc., Be.05 Heintz

Jackling, Daniel C., St.02 Jackling

Jacobsen, Henry, Da.01 Jacobsen

Jacobsen, Lydik S., St.01 Jacobsen

Jaffa, Myer E., Be.04; Be.05 Jaffa

Jaffé, George, Be.05 Jaffé

Jahns, Richard H., St.01 Jahns

James, Edwin, Be.05 James

Jenkins, Francis, Be.05 F. Jenkins

Jenkins, James, Be.05 J. Jenkins

Jenkins, Katharine, Be.05 K. Jenkins

Jenkins, Oliver, St.01 Jenkins

Jennings Radio Company, Be.05 Townsend

Jensen, Dilworth, Be.05 Jensen

Jepson, Willis, Be.05 Jepson

Jet Propulsion Laboratory, St.01 Siefert,

Johns Hopkins School of Medicine, SF.02 Rosencrantz

Johnson, Benjamin O., St.03 Johnson

Johnson, Francis, St.01 Johnson

Jones, Guy E., Be.05 G. Jones

Jones, Hardin, Be.05 H. Jones

Jordan, David Starr, Be.05 D. Jordan; St.01 D. Jordan | E. Jordan

Jordan, Eric, St.01 E. Jordan

Jordan, Peter, Be.05 P. Jordan

Jordanus, Be.05 Testi

Joslyn, Maynard, Be.05 Joslyn

Judah, Theodore, Be.05 Judah

Junck, Anita C., SF.02 Junck

Kaiser Wilhelm Gesellschaft, Be.05 E. Fischer

Kaiser, Henry J., Be.05 Six Companies

Kamen, Martin D., Be.05 Kamen

Kâmrân Shirazi, Medjumat-us-Sanai, St.06 Kâmrân Shirazi

Kantor, Harry, St.03 Kantor

Kaplan, Henry, St.01 Stanford University. Interviews

Karabadin Dehelvi, St.06 Karabadin Dehelvi

Keeler, James, SC.01 Keeler

Keen, Myra, St.01 Keen

Kellogg, Louise, Be.05 Alexander

Kennedy, Richard, St.03 Kennedy

Kerr, Clark, Be.02 Kerr | UCB, Chancellor

Kimble, George W., Be.05 Kimble

King, Clarence, Be.05 Farquhar

King, Ivan, Be.05 American Institute of Physics

Kingsley, John, Be.05 Kingsley

Kirkpatrick, Paul, St.01 Kirkpatrick; St.03 Kirkpatrick

Kittel, Charles, Be.05 Kittel

Kleiber, Max, Da.01 Kleiber

Klystrons, Be.05 Dodd | McCullough | Varian Associates | Woodyard | Terman; St.01 Hansen | Stanford University, Physics | Webster

Knight, Walter, Be.05 Purcell

Knudsen, Vern, Be.05 Knudsen

Knuth, Donald, St.01 Knuth | Stanford University, News

Kofoid, Charles, Be.05 Kofoid

Kompfner, Rudolf, St.01 Kompfner

Koshland, Daniel, Be.05 Koshland

Kotcher, Ezra, Be.05 Kotcher

Kreisel, George, St.01 Kreisel

Kroeber, Alfred L., Be.05 Kroeber | Kroeber-Quinn | Sparkman | Stewart

Kroeber-Quinn, Theodora, Be.05 Kroeber | Kroeber-Quinn

Krueger, Albert P., Be.05 A. Krueger

Krueger, Myron E., Be.05 M. Krueger

Kuala Lumpur, SF.02 Audy

Kurti, Nicholas, Be.05 Kurti

La Brea, Be.05 J. Merriam

Laboratories, Be.01; Be.02 Architects and Engineers; Be.05 George. *See also* entries under individual institutions

Lademan, Joseph V., St.03 Lademan

Laidlaw, Harry Hyde, Jr., Da.01 Laidlaw

Lammerts, Walter E., Be.05 Lammerts

Land reclamation, Be.05 Adams | Lowdermilk; Da.01 Holland Land Company; Sa.02 Natomas Company

Lane Hospital, St.01 Stanford University, President; St.06

Lane Medical Library, St.06

Lane, Levi Cooper, St.06 Halls of Aesculapius | Huntington | Lane | Lane Hospital | Swift

Lane Lectures, St.06 Huntington | Lane | Stanford University, Medical Center

Langelier, Wilfred F., Be.05 Langelier

Langley Porter Clinic, Be.05 Bowman

Langmuir, Irving, Be.05 Langmuir

Laplace, Pierre Simon, Be.03

Larmor, Joseph, Be.03

Lasers, St.01 Schawlow; St.02 USA vs. AT&T

Lawe, Annie, St.02 Lewis

Lawrence Berkeley Laboratory. *See* UCB

Lawrence Livermore Laboratory. *See* UCB

Lawrence, Ernest Orland, Be.01; Be.02 Childs | Lawrence | Radiation Laboratory | Research Corporation; Be.05 Fuller | Lawrence | G. N. Lewis. *See also* UCB, Lawrence Berkeley Laboratory

Lawson, Andrew, Be.05 A. Lawson | Palache | Seismological Society

Lawson, James S., Be.05 J. Lawson

LeBaron, Robert, St.03 LeBaron

LeConte, John, Be.05 LeConte | Popenoe

LeConte, John L., Be.05 J. L. LeConte,

LeConte, Joseph, Be.05 LeConte | Palache | Voy

Lévi-Strauss, Claude, Be.05 Kroeber

Lea, Isaac, Be.05 Sowerby

League for the Conservation of Public Health, Be.05 Sullivan

Lederberg, Joshua, St.01 ACME

Lee, John, Be.05 Lee

Lee, Russel Van Arsdale, St.06 Lee

Legge, Roy A., Be.05 Legge

Leighton, Robert, Be.05 American Institute of Physics

Lenard, Philipp, Be.03

Lenkurt Electric Company, Be.05 Appert

Lennette, Edwin, Be.05 Lennette

Lenzen, Victor, Be.02 Lenzen; Be.05 Lenzen

Leo XIII, Pope, Be.05 Bandelier

Leopold, A. Starker, Be.05 Leopold

Lesquereux, Leo, Be.05 Lesquereux

Leuschner, Armin Otto, Be.05 Leuschner; SC.01 Leuschner

Levi ben Gershon, Be.05 Levi ben Gershon

Lewis, Gilbert N., Be.05 Langmuir | G. N. Lewis

Lewis, James, St.02 Lewis

Lewis, Rubin M., Be.05 R. Lewis

Libby, Willard F., Be.05 Arnold | Libby

Liberty Farms Company, Da.01 Liberty Farms

Lick Observatory. *See* UC, Lick Observatory

Lick, James, SC.01

Lick Trust, SC.01 Floyd | Holden

Liddicote, Alfred R., Be.05 Eddy Tree Breeding

Linguistics, Be.05 Gifford | Kroeber | Lowie | Pinart | Powers

Linnaeus (Linné, Carl von), Be.05 Usinger

Lippincott, Donald, Be.05 Lippincott

Litton, Charles, Be.05 Buttner | Litton | McCullough

Livingood, John J., Be.05 Livingood

Loeb, Leonard, Be.05 Loeb

Loewner, Charles, St.01 Loewner

Lofgren, Edward J., Be.05 Lofgren

Logan, Thomas M., Be.05 G. Jones

Logic, Be.05 Bernstein | Tarski; St.01 Kreisel

Lombard, Henri, Be.05 Lombard

London, Ivan D., St.03 London

London School of Tropical Medicine, St.06 Reed

Lorentz, H. A., Be.03

Loring, Hubert S., St.06 Polio

Los Alamos Scientific Laboratory, Be.01; Be.02 Reagan | Underhill | Be.05 Bradbury | Brode | Dennes | Los Alamos | McMillan | Underhill | Woodyard; St.03 Agnew. See also Weapons testing

Louderback, George D., Be.05 Louderback | Seismological Society

Louisiana State University, Be.05 Jaffé

Lowdermilk, Walter, Be.04; Be.05 Lowdermilk

Lowell Observatory, St.02 See

Lowie, Robert, Be.05 Lowie

Luck, James, St.01 Luck

Madson, Ben A., Da.01 Madson

Magalotti, Lorenzo, Be.05 Accademia

Magendie, François, SF.02 J. Olmsted

Magnavox Corporation, Be.05 Albertus

Magnetrons, Be.05 Boot | Moore

Maindron, Ernest, Be.05 Maindron

Majors, Harry M., Be.05 Majors

Maker, Frank L., Be.05 Maker

Malacology, Be.05 S. Berry; St.01 Keen | Stanford University, Geology

Manhattan Engineer District, Be.01; Be.02 UCB, Radiation Laboratory; Brode | Calvin | Dennes | Lawrence | Libby | Los Alamos | Manhattan Engineer District | McMillan | Reynolds | Siri | C. S. Smith | Thornton | Underhill | St.03 Agnew | Blackwelder | Hiroshima-Nagasaki Publishing Committee | Rhodes | Young

Manley, John, Be.05 Los Alamos

Manson, Marsden, Be.05 Manson

Marg, Elwin, Be.05 Marg

Marine biology. See Biology, marine

Mark, J. Carson, Be.05 Los Alamos

Market Street Railroad, Sa.02 Moss

Marmont, George H., Be.05 Marmont

Marsh, Warner and Florence, Da.01 Marsh

Marshall, Lauriston Calvert, Be.05 L. Marshall

Marshall, Robert Bradford, Be.05 R. Marshall

Martens, W. L. F., Be.05 Martens

Marx, Charles David, St.01 C. Marx

Marx, Guido H., St.01 G. Marx

Marx, Karl, St.03 Cranberg

Mason, Herbert L., Be.05 Gifford

Massachusetts Institute of Technology, Be.05 Bowles | Hansen | Kittel | Massachusetts Institute of Technology | C. S. Smith; St.01 Hansen | Webster

Materia medica, St.06 Mir Muhammed | Muhammad Akbar | Muhammad Mu'min Husaini | Nedjibeddin es-Samerkandi

Mathematics, Be.02 Fuller | Thacher | Auhagen | Bélidor | Bernstein | Bošković | Cavaciocchi | Euclides | G. C. Evans | Morse | Pardies | Perry | Rabuel | Regnault | Rhiem | Rolle | Sefer mathematikah yashan w'nadir | Spira | A. Williams; St.01 Kreisel | Loewner; St.02 Miscelanea | Newton | Norman | See

——, applied, Be.05 G. C. Evans

——, education in, St.01 Forsythe

180

———, history of, Be.05 Cajori
———, and logic. See Logic
———, and physics, Be.05 Taub
———, practical, Be.05 Boodt,
Mathias, Mildred Esther, Be.05 Mathias
Matthes, François Emile, Be.05 Matthes
May, Bernice H., Be.05 May
Mayall, Nicholas, Be.05 American Institute of Physics
Mazzoni, Vincenzo, Be.05 Cavaciocchi
McCarthy, Edmund, Be.05 McCarthy
McCartney, Henry, Sa.01 McCartney
McCubbin, John, Da.01 McCubbin
McCulloch, Walter, Be.05 McCulloch
McCullough, Jack, Be.05 McCullough
McFarland, Frank, St.01 McFarland
McGaughey, Percy H., Be.05 McGaughey
McIntire, John, Sa.01 McIntire
McLaughlin, Donald H., Be.05 McLaughlin
McLaughlin, Major ———, Da.01 Edison
McMillan, Edwin M., Be.01; Be.02 Research Corporation; Be.05 McMillan
McMillan, Elsie Walford, Be.05 McMillan
McMillan, Howard G., SC.03
Mead, Elwood, Be.05 Mead
Mechanics, Be.05 Cerda | Chevalier | Cotes | Rabuel
Medical College of Pennsylvania, Be.05 Medical College
Medical College of the Pacific, St.06 Celsian Association | Medical College of the Pacific
Medical education, St.03 Butler; St.06. See also Nursing education
Medical physics, Be.01; Be.02 UCB, Donner Laboratory; Be.05 Anger | Born | Durbin | Gofman | Grendon | Hamilton | H. Jones | Lawrence | Myers | K. Scott | Siri | R. Stone | Van Dyke
Medicine, Be.01; Be.04; Be.05 Cutter Laboratories | Gregory | G. Jones | Poniatoff | Shumate | Storer; Da.01 Rowe; SF.02; St.01 ACME | Alway | Chandler | Clark | Mosher | Stanford University, Medical Center | Stanford University, Interviews | D. Wilbur | R. Wilbur; St.02 Medieval manuscripts | Scientists and science; St.03 Butler | Sams; St.06
———, Arabic, St.06 Abu-Hamid Muhammed | Al Ibrahimee | Al Kourashi | Gayet-ul-Beyan fil Tib
———, clinical, Be.05 Kaiser Permanente | Richter | Senn; Da.01 Craig; St.06. See also Hospitals
———, and electronics, Be.05 Marmont
———, Hindu, St.06 Muhammed Akber | Karabadin Dehelvi | Muhammad alad-din bin Hibat-Allah Adbzawari | Sarif Han
———, history of, SF.02 Harris | J. Olmsted; St.06 Barkan | Leeches | Spielmann | Stanford University, Medical Center | Van Patten
———, in Mexico, Be.05 Mexico, Archivo General; St.06 Segura
———, nuclear. See Medical physics
———, Persian, St.06 Kâmrân Shirazi | Muzaffer bin Muhammed ul-Husny | Tuhfat-ul-Mveminin der Tibb | Yusufi
———, research in, SF.02 California Society; St.06 California Society | Crawford | Polio | Rytand
———, in the South Pacific, SC.03
———, veterinary, Da.01 Hart
———, and war, SF.02 Brigham
———, and women, Be.05 Schlesinger Library | Medical College
Medico, Inc., St.06 Lee
Merriam, C. Hart, Be.05 Merriam | Stewart
Merriam, John, Be.05 Merriam
Metallurgy, Be.05 Butters | Christy | C. S. Smith; St.02 Jackling | Poncelet | Rabbitt
Metcalf, Woodbridge, Be.05 Metcalf
Meteorology, Da.01 Steinmetz; St.02 Medieval manuscripts
Metric system, Be.05 Drury | Kroeber
Metzger, H. Peter, St.03 Metzger
Mexia, Ynés, Be.05 Bracelin | Mexia

Mexico, Be.05 Cook | Mexico, Archivo General

Meyer, Karl F., Be.05 Meyer; SF.02 Harris

Microscopy, electron, Be.05 R. Williams

Microwaves, Be.05 Chodorow | Hansen | L. Marshall | MIT Radiation Laboratory | Vasseur | Whinnery; St.01. *See also* Electronics

Miller, Alden H., Be.05 A. Miller

Miller, Herman, Be.05 H. Miller

Miller, Loye, Be.05 L. Miller

Miller, Oliver, St.03 Miller

Miller, Stanley, Be.05 Arnold | S. Miller

Miller-Lux, Da.01 Miller-Lux

Mineralogy, Be.05 Palache | Yates | Yoakum; Sa.01; St.01 Park

Mining, Be.05 Butters | Christy | Dakin | Gulick | McCarthy | McLaughlin | Morris | Weeks; Da.01 Edison | Skaar Mining; Sa.02 Hall | Hendel | McCartney | Morse | National Bank | Natomas Company | Sutro | U.S. Land Office | Winchester; St.01 Park; St.02 Jackling; St.03 Rabbitt | Starr | Stuart

——, and engineering, Be.02 UCB, Mining Association; St.03 Hoover | Rabbitt | Starr

——, and mines, Sa.02 Carson Hill | McIntire | Mining companies | New Almaden Mine | New Idria

——, technology and, St.02 Hurst

Minkowski, Rudolph, Be.05 Minkowski

Mir Muhammed, Ilmi Tib, St.06 Mir Muhammed

Mirov, Nicholas, Be.05 Eddy Tree Breeding | Mirov

Missouri School of Mines, St.02 Jackling

Mitchell, Marion, Be.05 Mitchell

Monterey Institute of Foreign Studies, Da.01 Morrow

Montpellier, University of, St.02 Notebook

Moore, Arthur, St.01 A. Moore

Moore, George, St.01 G. Moore

Moore, Joseph, SC.01 Moore

Moore, Norman, Be.05 Moore

Morales, José, Be.05 Sturtevant

More, Henry, Be.05 Newton

Morgan, Agnes Fay, Be.05 Morgan

Morris, Fred, Be.05 Morris

Morris, Samuel, St.01 Morris

Morrow, Dwight, Da.01 Morrow

Morse, A. P., Be.05 Morse

Morse, Ephraim W., Sa.01 Morse

Morton, William, St.06 Morton | Warren

Moseley, H. G. J., Be.03

Mosher, Clelia, St.01 Mosher; St.06 Mosher

Moss, Joseph, Sa.01 Moss

Motion picture technology, St.02 Exactus Photo-Film Corporation

Mount Diablo base and meridian, Be.05 Ransom

Mount Hamilton. *See* UC, Lick Observatory

Mount Whitney, Be.05 Farquhar | Legge

Mount Wilson Observatory, Be.05 Hale

Moustafa ibn Halil us-Safiee, Kitab-ul-Edjsad, St.06 Moustafa ibn Halil us-Safiee

Moyer, Burton J., Be.05 Moyer

Mrak, Emil, Be.05 Mrak; Da.01 Mrak

Muhammad Akbar, Karabudin Kaderee, St.06 Muhammad Akbar

Muhammad Mu'min Husaini, Tufet-ul-Muminin, St.06 Muhammad Mu'min Husaini

Muhammad alad-din bin Hibat-Allah Adbzawari, St.06 Muhammad alad-din bin Hibat-Allah Adbzawari

Muhammed Akber, Mudjerrebat-i-Akberee, St.06 Muhammed Akber

Muir, John, Be.05 H. M. Evans | Muir

Munshi-Mazhar-al-Din, St.06 Munshi-Mazhar-al-Din

Munson, Clara S., St.06 Alvarez

Museums, Be.02 UCB, Anthropology | UCB, Vertebrate Zoology; Be.05 Palache | A. Stone

——, science, Be.02 Lawrence Hall of Science; Be.05 Cooper | Exploratorium | Storer | Taylor | White

Muzaffer bin Muhammed ul-Hvsny, Tibb-i-Shiffa'ee, St.06 Muzaffer bin Muhammed ul-Hvsny

Myers, William G., Be.05 Myers

National Academy of Sciences, Be.05 Hale; SC.03; St.01 Schiff. *See also* entries for individual members

National Aeronautics and Space Administration, Be.05 COMSAT; St.03 Beggs

National Bank of D. O. Mills & Company, Sa.01 National Bank

National Canners' Association, St.06 Dickson

National Institutes of Health, Be.05 Yerushalmy

National Research Council, Be.05 Hale | National Research Council

National Resources Committee, St.01 Morris

Natomas Company, Sa.01 Natomas Company

Natural history, Be.05 Botta | Bryant | Emerson | C. Merriam | Muir | A. Stone | Yates; St.01 Wiggins; St.06 Ghigath-al-Din Ali al Husayn al'Isfah in | Shal mardan al Mustawf

Natural philosophy, Be.05 Bošković | St.02 Newton

Navigation, Be.05 Davidson | Demarcaciones | Drawings | South Sea Waggoner | Thevet | Thompson

Nederlandsche Artsenkamer, St.03 Nederlandsche Artsenkamer

Nedjibeddin es-Samerkandi, St.06 Nedjibeddin es-Samerkandi

Needham, Paul R., Be.05 Needham

Nelli, Giovanni Battista Clemente, Be.05 Galileo

Nelson, Dewitt, Be.05 Nelson

Nelson, Oscar, Da.01 Nelson

Neubauer, Loren W., Da.01 Neubauer

Neumann, John von, Be.05 Taub

Neurology, St.06 Cooper Medical College

New Almaden Mine, Sa.01 New Almaden

New Idria Mining Company, Sa.01 New Idria New South Wales Medical Board, St.03 New South Wales

New York University, St.06 Warren

Newton, Sir Isaac, Be.05 Cajori | Newton; St.02 Newton

Neyman, Jerzy, Be.05 Neyman

Nobel Prize, Be.02 Alvarez | Bohr | Chamberlain | Einstein | E. Fischer | Giauque | Lawrence | Libby | E. McMillan | Seaborg | Segrè | Stanley | Stern; St.01 Berg | Bloch | Schawlow | Shockley

Norman, Haskell, St.02 Norman

Northern Transcontinental Survey, Be.05 Hilgard

Northwestern Pacific Railway, Be.05 Northwestern Pacific Railway

Norton, Charles, Be.05 Norton

Notgemeinschaft der deutschen Wissenschaft, St.03 Deutsche Forschungsgemeinschaft

Nuclear chemistry. *See* Chemistry, nuclear

Nuclear medicine, Be.05 Anger | Born | Gofman | Grendon | Hamilton | Myers

Nuclear physics. *See* Physics, nuclear

Nuclear science, Be.01; Be.02 Childs | Lawrence; Be.05 Fidler | Seaborg | C. S. Smith. *See also* Physics; Chemistry; Medicine

Nuclear weapons, Be.05 Lawrence | Los Alamos | Manhattan Engineer District | York; St.03 Hiroshima-Nagasaki Publishing Committee | Lademan | LeBaron | Pash. *See also* Manhattan Engineer District; Radiation; Weapons testing

Nursing, Be.02 UCB, Public Health; SF.02 Sigma Theta Tau; St.06 De Forest | Gould

———, education for, St.06 Gould | Stanford University, Nursing School

Nutrition, Be.04 Jaffa; Be.05 California Public Health | Jaffa | Morgan. *See also* Food science

Oak Ridge National Laboratory, Be.05 Reynolds

Observatories, astronomical, Be.05 Galloway | Hale | Hill | Leuschner | Palmer | Struve; SC.01. *See also* Astronomy

Obstetrics and gynecology, SF.02 Arnot; St.06 San Francisco Maternity | Stanford Clinics Auxiliary

Oceanography, SC.03; St.02 Ricketts; St.03 Intergovernmental Oceanographic Commission. *See also* Biology, marine

Œnology and viticulture, Be.05 Amerine | Cruess | Joslyn | Tchelistcheff | Da.01 Amerine | Audio-visual | Cruess | Guymon | Portuguese wine pamphlets | Vermouth Project | Wine newsletters | Wines and vines. *See also* Wines and wineries; Viticulture

Ohio State University: Be.05 Myers

Oil industry, St.03 Elliott

Oldroyd, Mrs. Ida Shephard, St.02 Oldroyd

Olmo, Harold P., Be.05 Olmo

Olmsted, Evangeline Harris, SF.02 E. Olmsted

Olmsted, James, SF.02 E.Olmsted | J. Olmsted

Oncology, St.01 Kaplan; St.06 Gunn

Operations research, Be.05 Churchman | Crandall

Ophthamology, St.06 Barkan | Bettman | Cooper Medical College | Duke-Elder

Ophüls, William, St.06 Ophüls, Reed

Oppenheimer, Frank, Be.05 American Institute of Physics | Exploratorium

Oppenheimer, J. Robert, Be.05 Dennes. *See also* Manhattan Engineer District

Optics, Be.05 Marg; St.01 Kompfner

Optometry, Be.05 Marg; St.01 Kompfner

Oral history. *See* appended list of oral history interviews

Oregon State University, Be.05 McCulloch

Ornithology, Be.05 Allen | Audubon Association | Barlow | Bishop | Bryant | Chambers | Cooper Ornithological Society | Grayson | Grinnell | Leopold | A. Miller | L. Miller | Tyler; St.02 Seale

Osler, Sir William, SF.02 Rosencrantz

Osteology, St.06 Lane

Osteopathy, St.06 Bone and joint deformities

Oxford University, Be.05 Kurti

Pacific Biological Laboratories, St.02 Ricketts,

Pacific Coast Surgical Association, SF.02 Pacific Coast Surgical

Pacific Gas and Electric Company, Be.05 Ellis | Hedgpeth | Pollard; Sa.02 Moss. *See also* Power

Pacific Geographic Society, St.02 Pacific Geographic Society

Pacific Ocean, Be.05 Botta; SC.03; St.01 Vanderbilt Foundation; St.02 Pacific Geographic Society

Packard, Walter, Be.05 Packard

Packard, David, St.02 USA vs. AT&T; St.03 Packard. *See also* Hewlett-Packard

Page, Stanley H., St.02 Hopkins

Palache, Charles, Be.05 Palache

Paleobotany, Be.05 Chaney

Paleontology, Be.05 Alexander | Baldwin | Camp | Chaney | Deans | M. Hildebrand | J. Merriam | L. Miller | Yates

Palmer, Harold, Be.05 Palmer

Palo Alto Hospital, St.01 Stanford University, Medical Center

Palo Alto Medical Clinic, St.01 Clark

Palomar Observatory, Be.05 Hale

Panama Canal, St.01 J. Branner

Panofsky, Wolfgang, St.04 Director | Scientific records Pardies, Ignace Gaston, Be.05 Pardies

Paris, science in, Be.05 Some account

Park, Charles F., Jr., St.01 Park

Parry, Charles, Be.05 Parry

Pash, Boris T., St.03 Pash

Patents, Be.02 UCB, Patent Office; Be.05 Fuller | Heintz | Lippincott | Maker | R. Williams. *See also* Inventions and inventors

Pathology, Be.05 Gardner | Meyer | Smith; St.02 Notebook; St.06 Cooper Medical College | Ophüls | Pathologia;

Patt, Harvey, SF.02 Patt

Pauly, ———, Be.05 Pauly

Payne-Gaposchkin, Cecilia, Be.05 Dieter

Peckham, Stephen, St.02 Peckham

Peirce, Benjamin, Be.05 Darwin,

Peirce, Charles, Be.05 Peirce

Peirce, George, St.01 Peirce

184

Perham, Douglas, Be.05 Perham

Perry, Newel, Be.05 Perry

Peters, T. K., St.02 Exactus Photo-Film Corporation

Peterson, Maurice L., Da.01 Peterson

Petroleum, Be.05 Louderback. *See also* Engineering, petroleum

Pharmaceuticals and pharmacology, Be.05 Cutter Laboratories | Schneider; SF.02 Burbridge | Riegelman; St.06 Crawford

Phelps, Ralph L., Be.05 Butters

Philosophy, Be.05 De scientiarum | Dennes; St.02 Fauche-Borel | Medieval manuscripts

Photography and photographs: Be.05 Einarsson; SC.01 Photographic collection

Photosynthesis, Be.05 Calvin | Kamen

Physical sciences, St.01 Stanford University, Physical Sciences

Physics, Be.02 Childs | Lawrence | Lenzen | Teller | UCB, Physics; Be.03; Be.05 Alvarez | American Institute of Physics | Auhagen | Birge | Bošković | Brode | Chamberlain | Cotes | Hansen | Helmholz | Jaffé | F. Jenkins | Kittel | Knudsen | Langmuir | Lawrence | LeConte | Litton | Lofgren | L. Marshall | Massachusetts Institute of Technology | Natonal Research Council | Slate | White; St.01 D. Jordan | Kirkpatrick | Schawlow | Stanford University, Physics | Stanford University, Interviews | Stanford University, Physical Sciences | Webster; St.02 Franklin | Scientists and science | See; St.03 Billings | Cranberg | Gilinsky | Kirkpatrick | Ruark | Teller | Shal mardan al Mustawf. *See also* Mechanics; Natural philosophy

———, applied, St.01 Kompfner

———, at Berkeley, Be.01; Be.02; Be.03; Be.05 Birge

———, biophysics and spectrography, SF.02 Strait

———, electricity and magnetism, Be.05 Purcell

———, gaseous discharge, Be.05 Loeb

———, high-energy, Be.05 International Union; St.04

———, low-temperature, Be.05 Giauque | Kurti

———, mathematical, Be.05 Taub

———, nuclear, Be.01; Be.05 Goldhaber | Henderson | Lawrence | Livingood | E. McMillan | Moyer | Segrè | Thornton | York; St.01 Bloch | Schiff; St.04

———, particle, St.04 High-energy physics. *See also* Accelerators; Cyclotrons

———, and philosophy, Be.05 Lenzen,

———, quantum, Be.03; Be.05 Archive | Bohr | Stern | Wise; St.01 Schiff; St.03 Teller Physiology, Be.05 Cook | Durbin | H. M. Evans | H. Jones | Marmont | Torrey | Wood; SF.02 J. Olmsted; St.01 Jenkins | R. Wilbur | Weymouth; St.02 Fauche-Borel; St.06 Doremus | Drinker | Hunter | R. Wilbur

———, animal, Da.01 Kleiber

———, plant, Be.05 Mirov; St.01 Peirce

Piccioni, Oreste, Be.05 Segrè

Pickelner, S. B., Be.05 Dieter

Pinart, Alphonse, Be.05 Pinart

Pinger, Roland W., Be.05 Pinger

Pinner, B., Be.05 Kroeber

Planck, Max, Be.03

Plummer, Richard H., St.06 Plummer

Pneumatics, Be.05 Cotes

Polio, Be.05 Cutter Laboratories; St.06 Polio research

Pollard, E. C., Be.05 Hansen | Pollard

Pomology, Da.01 Corti | Pomology

Poncelet, Eugene F., St.03 Poncelet

Poniatoff, Alexander, Be.05 Poniatoff

Pont, Antonino Marcó del, St.06 Pont

Popenoe, Edwin A., Be.05 Popenoe

Poppe, Nikolai N., St.01 Poppe

Population studies, St.01 Ehrlich

Porter, Robert Langley, Be.05 Porter

Portsmouth Collection, Be.05 Newton

Powell, John Wesley, Be.05 Powers | Sparkman

Powell, Richard, Be.05 Powell

Power, Robert, Da.01 Power/Drake

Power, electric, Be.05 Ellis | Northwestern Pacific Railway | Powell | Statewide Conference; Sa.02 Moss

Power, nuclear, Be.01; Be.05 Fidler | Hedgpeth | Seaborg | Sierra Club | C. S. Smith | Untermyer; St.01 Stephen; St.03 Gilinsky | Kennedy | LeBaron | Metzger | Ray

Power, water, Be.05 Gaytes; Be.06. *See also* Water resources

Powers, Stephen, Be.05 Powers

Pratt, Haraden, Be.05 Pratt

Prentice, Arthur, St.06 Prentice

Prince, Helen D., Be.05 Dieter

Princeton University, Be.05 Arnold

Prohibition and repeal, Da.01 California wineries

Project Sherwood, Be.05 York

Pseudo-science, St.03 Cranberg

Psychiatry, Be.05 Bowman | Porter; St.02 Berillon

Psychoanalysis, SF.02 Berne | Fraiberg
Psychology, Be.05 J. Holmes | Stratton; St.01 Farnsworth | L. Terman | Sears | Stanford University, Hypnosis Laboratory | Stanford University, Psychology Department | Stolz | Zimbardo; St.02 Ritter; St.03 London

Psychopharmacology, SF.02 Burbridge

Public health, Be.02 UCB, Public Health; Be.05 Arnstein | California Public Health | Kaiser Permanente | Beattie | Bolt | Chance | G. Jones | Meyer | Porter | Schlesinger Library | C. E. Smith | Sullivan | Warren | Wyckoff | Yerushalmy; Da.01 Storer; Sa.01; SF.02 Junck; SC.03; St.01 Mosher; St.06 Sanitary surveys

Purcell, Edward, Be.05 Purcell

Putnam, John, Be.05 Putnam

Putnam, Katharine, Be.05 Muir

Pyramid Project, Be.05 Alvarez

Quantum physics. *See* Physics, quantum,

Quesnel, Julius, Be.05 Baldwin

Radio Corporation of America (RCA), Be.05 RCA

Rabbitt, James, St.03 Rabbitt

Rabuel, Claude, Be.05 Rabuel

Radar, Be.05 Boot | Guerlac | Massachusetts Institute of Technology | Terman | Vasseur; St.01 F. Terman | Hansen; St.03 Miller

Radcliffe College, Be.05 Schlesinger Library

Radiation Laboratory. *See* UCB, Lawrence Berkeley Laboratory

Radiation, effects of, Be.05 Hamilton | Hedgpeth | H. Jones | Lawrence | Moyer | K. Scott; Da.01 Born | Crafts; St.01 Stephen; St.03 Sams. *See also* Medical physics; Medicine, nuclear

Radio and radio engineering, Be.05 Buttner | Chadbourne | Clement | Dollar | Fisher | Fuller | Griffith | Litton | H. Miller | RCA | Scofield | Simon | Sinclair | H. Stone | Vasseur | White; St.02 Elwell

———, and business, Be.05 Lippincott | Townsend

Radio, history of, Be.05 Ashton | Pratt | Vasseur

Radioastronomy, Be.05 Dieter

Radiobiology, SF.02 Patt

Radiocarbon, Be.05 Arnold | Calvin | Kamen

Radiology, Be.05 R. Stone; St.01 Kaplan

Radiotelephone, St.03 Gilman

Railroads, Be.05 Randall; Sa.02 McCartney | Moss | National Bank | Sutro; St.02 Hopkins | Hyde; St.03 Darling | Emerson | Hoskin | Johnson | Stevens | U.S. Advisory Commission

———, and electrification, Be.05 Northwestern Pacific Railway | Pollard

Rambo, William R., St.01 Rambo

Randall, Henry, Be.05 Randall

Ransom, Leander, Be.05 Ransom

Rawn, A. M., Be.05 Rawn

Ray, Dixy Lee, St.03 Ray

Reagan, Ronald, Be.02 Reagan

Reed, Alfred, St.06 Reed

Regnault, Henri, Be.05 Regnault

Reichsanzeiger, St.06 Medicinalia et chirurgica

Relativity, St.01 Schiff. *See also* Physics
Religion and science, Be.05 Lammerts
Research Corporation, Be.02 Research Corporation
Reynolds, Wallace B., Be.05 Reynolds
Rhiem, Johann, Be.05 Rhiem
Rhodes, Fred, St.03 Rhodes
Rhodin, Carl J., Be.05 Rhodin
Richardson, Sir Benjamin Ward, Da.01 Richardson
Richardson, George, St.01 Richardson
Richter, Clemens M., Be.05 Richter
Richter, Max Clemens, Da.01 Richter
Ricketts, Edward F., St.02 Ricketts
Riegelman, Sidney, SF.02 Riegelman
Righter, Francis I., Be.05 Eddy Tree Breeding
Rising, Willard, Be.05 Rising
Ritter, [August] Heinrich, St.02 Ritter
Ritter, William, Be.05 Ritter
Rivers, John, Be.05 Treadwell
Rix, Edward, Be.05 Rix
Rixford, Emmet, St.06 Rixford
Roberts, Eugene L., Be.05 Roberts
Rockefeller Foundation, St.03 Gunn
Rocketry, St.01 Siefert
Rocky Mountain Biological Laboratory, St.01 Ehrlich
Rocky Mountain news, St.03 Metzger
Rolle, Michel, Be.05 Rolle
Rosenberg Foundation, Be.05 Chance
Rosencrantz, Esther, SF.02 Rosencrantz
Rowe, Albert H., Da.01 Rowe
Rowell, Chester, Be.05 Sullivan
Rowntree, Lester, Be.05 Rowntree | Mathias
Royal Botanic Garden, Be.05 Burke
Royal Navy Scientific Service, Be.05 Boot
Royal Society of London, Be.05 Ulloa
Ruark, Arthur Edward, St.03 Ruark
Ruben, Samuel, Be.05 Kamen

Russia, Be.05 M. Sauer. *See also* Soviet science
Russian Railway Service Corps, St.03 Emerson | Hoskin | Johnson
Rutherford, Ernest, Be.03
Ryan, Harris J., Be.05 National Research Council; St.01 Ryan | Stanford University, Ryan High Voltage Laboratory
Ryerson, Knowles A., Da.01 Ryerson; SC.03
Rytand, David A., St.06 Rytand
SRI International, St.05
Saint Mary's Hospital, San Francisco, SF.02 Arnot
Sams, Crawford, St.03 Sams
San Cristóbal Observatory, Be.05 Palmer
San Francisco, Be.05 Ellis; Sa.02 Sutro
San Francisco City and County Hospital, St.06 San Francisco City and County Hospital
San Francisco County Hospital, St.06 Rixford
San Francisco County Medical Society, St.06 San Francisco County Medical Society
San Francisco Maternity, St.06 San Francisco Maternity
San Francisco Medical Society, St.06 Whitney
San Francisco Medico-Chirurgical Society, St.06 Lane
San Quentin Prison, St.06 Stanley
Sapir, Edward, Be.05 Gifford
Sarif Han, Iladj-ul-Amradz, St.06 Sarif Han
Sauer, Carl O., Be.05 Cook | C. Sauer
Sauer, Martin, Be.05 M. Sauer
Save-the-Redwoods League, Be.05 Fritz
Schawlow, Arthur L., St.01 Schawlow | Stanford University, News and Publications; St.02 USA vs. AT&T
Schenk, Hubert G., St.01 Stanford University, Geology
Scheuring, Ann, Da.01 Scheuring
Schiff, Leonard I., St.01 Schiff
Schlesinger Library, Radcliffe College, Be.05 Schlesinger Library
Schmalzing, Heinrich, St.06 Medicinalia et chirurgica

Schmidt, Alexis von, Be.05 Mitchell

Schneider, Albert, Be.05 Schneider

School of Tropical Medicine, Bombay, Be.05 Kofoid

Schreiber, Raemer, Be.05 Los Alamos | Schreiber

Schulz, Helmut W., Be.05 Schulz

Schumacher, Paul, Be.05 Schumacher

Schuster, Arthur, Be.03

Schwarzschild, Karl, Be.03

Schwerdt, Carlton, St.06 Polio

Schwettman, Alan, St.01 Stanford University, Interviews

Science Service, Be.05 Ritter

Science education, Be.02 J. Hildebrand | Lawrence Hall of Science | Lenzen | Student notes; Be.05 Exploratorium | May; St.01 Forsythe

Science policy, St.03 Billings | Metzger. See also U.S. Atomic Energy Commission

Scintillation camera, Be.05 Anger

Scofield, Philip, Be.05 Scofield

Scott, Flora, Be.05 F. Scott

Scott, Kenneth, Be.05 K. Scott

Scripps Institution, Be.05 Fox | Ritter

Seaborg, Glenn, Be.02 UCB, Chancellor; Be.05 Seaborg

Seale, Alvin, St.02 Seale

Sealy, S., Be.05 Hansen; St.01 Hansen

Sears, Robert, St.01 Sears

See, T. J. J., St.02 Brasch | See

Segrè, Emilio, Be.05 Segrè

Segura, Ignacio, St.06 Segura

Seismological Society of America, Be.05 Seismological Society

Seismology, Be.05 Byerly | A. Lawson | Seismological Society; St.01 J. Branner | Earthquake collection | Jacobsen

Semino, Angelo, St.01 Semino

Senn, Milton, Be.05 Senn

Servetus, Michael, St.06 Lane

Sessler, Andrew, Be.01

Setchell, William, Be.05 Setchell

Sexuality, St.01 Mosher

Shal mardan al Mustawf, Nvzhat Nameh-i-Alai, St.06 Shal mardan al Mustawf

Shane, C. Donald, Be.05 K. Campbell | Shane; SC.01 Shane. See also UC, Lick Observatory

Shane, Mary Lea, Be.05 Shane; SC.01. See also UC, Lick Observatory

Sharpe, Bartholomew, Be.05 South Sea Waggoner

Sherman, Anne, Be.05 Smathers

Shields, Peter J., Da.01 Shields

Shockley, William B., St.01 Shockley

Shumate, C. Albert, Be.05 Shumate

Si hat ad-din bin Abd al-Karim, St.06 Si hat ad-din bin Abd al-Karim

Siefert, Howard, St.01 Siefert

Sierra Club, Be.05 Clark | J. Hildebrand | Sierra Club | Siri. See also Conservation

Sight impairment, Be.05 Perry. See also Optometry; Ophthamology

Sigma Theta Tau, SF.02 Sigma Theta Tau

Silicon Valley, Be.05 Terman | Varian Associates; St.01 F. Terman

Silver, Samuel, Be.05 Silver

Simon, Emil, Be.05 Simon

Simpson, Edmund E., Da.01 Botanical prints

Simpson, Miriam, Be.05 H. M. Evans

Sinclair, Donald, Be.05 Sinclair

Siri, William, Be.05 Siri

Six Companies, Inc., Be.05 Six Companies

Sizer, Frank L., Be.05 Sizer

Skaar Mining Company, Da.01 Skaar Mining

Skilling, Hugh, St.01 Skilling

Slate, Frederick, Be.05 Slate

Smathers, James, Be.05 Smathers

Smith, Charles, Be.05 C. E. Smith

Smith, Cyril, Be.05 C. S. Smith

Smith, Jedediah, Be.05 Cremer

Smith, Jeremiah, Be.05 J. Smith

Smith, Ralph, Be.05 R. Smith

Smithsonian Institution, Be.05 Cooper | Legge | Mexía

Snyder, C. C., Be.05 Pollard
Social welfare, Be.05 Arnstein
Society of American Foresters, Be.05 Metcalf
Society of Jesus, Be.05 Bošković
Soil science, Be.05 Blaney | Hilgard | S. Tibbetts; Da.01
South Pacific Archives, SC.03
South Pacific Collection, SC.03
South Pacific Commission, SC.03
South Sea Waggoner, Be.05 South Sea Waggoner
Southern Pacific Railroad, Be.05 Randall | St.02 Hyde
Sovary, Lily, SF.02 Sovary; St.06 Sovary
Soviet Union, science and engineering in, St.03 London | Luck. *See also* U.S. Advisory Commission; Railroads
Sowerby, Miss, Be.05 Sowerby
Space exploration and science, Be.05 COMSAT | L. Marshall | Silver; St.03 Beggs | Weakley. *See also* U.S. National Aeronautics and Space Administration
Sparkman, Philip, Be.05 Sparkman
Sperry Gyroscope Company, Be.05 Dodd | Be.05 Woodyard
Sperry Rand Corporation, St.01 Hansen. *See also* Klystrons
Spielmann, Marion H., St.06 Spielmann
Spier, Anna, Be.05 A. Spier
Spier, Leslie, Be.05 L. Spier
Spier, Robert, Be.05 R. Spier
Spieth, Herman, Be.05 Spieth
Spink, Henry, Be.05 Spink
Spinrad, Myron, Be.05 American Institute of Physics
Spira, Robert, Be.05 Spira
Sproul, Robert, Be.02 Sproul
Standard Oil Company of New York, and China, Be.05 Louderback
Standard Oil Company of California, Be.05 Crandall | Putnam
Stanford Clinics Auxiliary, St.06 Stanford Clinics Auxiliary
Stanford Convalescent Home, St.06 Faber

Stanford Linear Accelerator Center (SLAC), St.04
Stanford University, St.01–06 Stanford University; Be.05 Fuller | Hansen | D. Jordan | H. Miller | Terman | Varian Associates | Woodyard; SF.02 Harris
Stanley, Leo Leonidas, St.06 Stanley
Stanley, Wendell Meredith, Be.05 Stanley
Starr, Clarence T., St.03 Starr
Statewide Conference on Undergrounding Utilities, Be.05 Statewide Conference
Statics, Be.05 Pardies. *See also* Mechanics
Statistics, Be.05 Churchman | Neyman
Steam car, Be.05 Doble
Stebbins, George, Da.01 Stebbins
Steinmetz, Andrew, Da.01 Steinmetz
Stephen, Mark, St.01 Stephen
Stepanov, A. I., St.03 Stepanov
Stern, Curt, Be.05 R. Goldschmidt
Stern, Otto, Be.05 O. Stern
Stevens, John, St.03 Stevens
Stewart, George W., Be.05 Stewart
Stolz, Lois, St.01 Stolz
Stone, Andrew, Be.05 A. Stone
Stone, Herbert, Be.05 H. Stone
Stone, Robert, Be.05 R. Stone
Storer, Ruth R., Be.05 Storer
Storer, Tracy, Be.05 Storer; Da.01 Storer
Strait, Louis A., SF.02 Strait
Stratton, George, Be.05 Stratton
Strauss, Joseph, St.02 Strauss
Strong, Edward, Be.02 UCB, Chancellor
Struve, Otto, Be.05 Struve
Stuart, Charles, St.03 Stuart
Sturtevant, William C., Be.05 Sturtevant
Submarine detection, Be.05 Knudsen
Sugar refining, Da.01 Nelson
Sullivan, Celestine J., Be.05 Sullivan; St.06 Sullivan
Sultan Ali tabib al Hurâsâni, Dustur ul-Iladji, St.06 Sultan Ali tabib al Hurâsâni
Surgery, Be.05 Gerbode | James | R. Lewis | Wood; SF.02 Brigham | Dunphy |

Inman | Pacific Coast Surgical Association; Surgery, St.01 Chandler; St.06 Cooper Medical College | Duke-Elder | Medicinalia et chirurgica

Surveys and surveying, Be.05 Ashburner | Aughey | Brewer | California Geological Survey | Canada Geological Survey | Cooper | Elliott | Farquhar | Gilman | Hilgard | Hoffmann | J. L. LeConte | Kimble | Martens | C. H. Merriam | Peirce | Ransom; Sa.02 Hendel | McCartney | Mining companies. *See also* California Geological Survey; Cartography; Exploration; Geology; Geography; Railroads; U.S. Coast and Geodetic Survey; U.S. Geological Survey

Susskind, Charles, Be.05 Chodorow

Sutro Tunnel, Sa.02 Sutro

Sutro, Adolph, Be.04; Sa.01 Sutro; St.02 Sutro

Swain, Robert Eckles, St.01 Swain

Sweasey Powers, W. J., St.06 Sweasey Powers

Swift, Sidney, St.06 Swift

Taiwan, St.03 Billings

Tarski, Alfred, Be.05 Tarski

Taub, Abraham H., Be.05 Taub

Taylor, Walter P., Be.05 Taylor

Tchelistcheff, André, Be.05 Tchelistcheff

Technology, Be.02 UCB, Appropriate Technology; Be.05 Gilfillan; Sa.01; SC.03; St.02 Hurst; St.03 Billings; St.06 Kâmrân Shirazi

———, agricultural, Da.01 Agricultural technology | Cove/Bakken | Higgins | Liberty Farms | Photographs | Walker; Sa.01

———, and business, Be.05 Coast Manufacturing | Crown Zellerbach | Eitel-McCullough | O. Fischer | Heintz | Litton | Poniatoff

———, food, Be.05 Mrak; Da.01 Pomology; Sa.02 Hiller

———, nuclear, Be.05 Untermyer. *See also* Power, nuclear; U.S. Atomic Energy Commission

———, sound recording, Be.05 Poniatoff

———, in the Soviet Union, St.03 Stepanov

———, and winemaking. *See* Œnology

Telecommunications, St.02 USA vs. AT&T

Telegraphy, Be.05 Field; Sa.02 Moss; St.02 Hyde

Telescopes. *See* Astronomy; Observatories

Television, Be.05 Lippincott | Vasseur

Teller, Edward, Be.02 Teller; St.01 Stanford University, News and Publications; St.03 Teller

Terman, Frederick, Be.05 Terman; St.01 Bourguin | Stanford University, News and Publications | F. Terman | L. Terman

Terman, Lewis M., St.01 Terman

Testi, Cesare, Be.05 Testi

Thacher, Edwin, Be.02 Thacher

Thevet, André, Be.05 Thevet

Thompson, Robert, Be.05 Thompson

Thomson, Sir John A., Be.05 Thomson

Thornberg, Royden, Be.05 Perham

Thornton, Robert L., Be.05 Thornton

Three Mile Island, St.01 Stephen; St.03 Gilinsky. *See also* Power, nuclear

Throop College of Technology. *See* California Institute of Technology

Tibbetts, Frederick, Be.05 F. Tibbetts

Tibbetts, Sydney A., Be.05 S. Tibbetts

Tokyo Daigaku, St.03 Tokyo Daigaku Statement

Topography, Be.05 Matthes

Torrey, Harry, Be.05 Torrey

Townes, Charles, St.02 USA vs. AT&T

Townsend, Calvin, Be.05 Townsend

Transactional Analysis Institute, San Francisco, SF.02 Berne

Transportation, St.02 Hopkins Transportation Collection V. *See also* Railroads

Treadwell, George A., Be.05 Treadwell

Tuhfat-ul-Mveminin der Tibb, St.06 Tuhfat-ul-Mveminin der Tibb

Turner, Francis, Be.02 Turner; Be.05 Turner

Tuttle, Albert H., Be.05 Tuttle
Tyler, John G., Be.05 Tyler
Typewriters, electric, Be.05 Smathers

U.S. Advisory Commission of Railway Experts to Russia, St.03 Stevens | U.S. Advisory Commission

U.S. Air Force, St.01 Schiff; St.03 Miller | Young

U.S. Army, St.01 Webster; St.03 Pash | Sams; St.06 Wesson

U.S. Atomic Energy Commission, Be.01; Be.05 Fidler | Lawrence | Libby | Los Alamos | Seaborg | C. S. Smith; St.03 LeBaron | Ray | Ruark. *See also* Manhattan Engineer District; U.S. Department of Energy

U.S. Biological Survey, Be.05 C. Merriam

U.S. Bureau of Alcohol, Tobacco, and Firearms, Da.01 California wineries

U.S. Bureau of American Ethnology, Be.05 Hodge | Sparkman

U.S. Coast and Geodetic Survey, Be.05 Davidson | Elliott | Hilgard | J. Lawson | Peirce. *See also* Surveys

U.S. Commercial Company, SC.03

U.S. Department of Defense, Be.02 Underhill; St.03 LeBaron

U.S. Department of Energy, Be.01; St.04 Department of Energy

U.S. Department of the Interior Department, St.01 R. Wilbur

U.S. Department of Labor, St.03 Kantor

U.S. Engineer's Office, Be.05 Ransom

U.S. Federal Records Center, ii; St.04 Department of Energy

U.S. Forest Service, Be.05 Coffman | Mirov | Nelson

U.S. Fuel Administration, St.03 Stuart

U.S. Geological Survey, Be.05 Cooper | Hodge | Jahns | A. Lawson | Matthes | McLaughlin

U.S. Indian Claims Commission, Be.05 Cook. *See also* Indians

U.S. Internal Revenue Service, Da.01 Federal wine bottle labels

U.S. International Cooperation Administration, St.03 Kirkpatrick

U.S. Land Office, Sacramento, Sa.01 U.S. Land Office

U.S. Library of Congress, Be.05 Langmuir; St.02 Brasch

U.S. Medical Corps, St.03 Butler

U.S. National Advisory Commission on Food and Fiber, Da.01 Decker

U.S. National Aeronautics and Space Administration, St.01 Schiff; St.03 Weakley. *See also* Space

U.S. National Archives, Be.05 Cook | Guerlac | Manhattan Engineer District | Powers

U.S. National Defense Research Committee, Be.05 G. C. Evans | Neyman; St.01 F. Terman. *See also* War and science

U.S. National Park Service, Be.05 Bryant | Coffman | R. Marshall. *See also* Conservation

U.S. Navy, Be.05 Kittel | Loeb | U.S. Navy | White | Wood; St.03 Lademan

———, Naval Postgraduate School, Monterey, Be.05 Marmont

U.S. Nuclear Regulatory Commission, St.01 Stephen; St.03 Gilinsky | Kennedy. *See also* Power, nuclear

U.S. Public Health Service, Be.05 Yerushalmy

U.S. Revenue Marine, Be.05 Churchill

U.S. Rubber Company, Be.05 Villars

U.S. Signal Corps, Be.05 Lippincott

U.S. Veterans' Administration Hospital, San Francisco, SF.02 Dunphy

U.S. Work Projects Administration, Be.02 U.S. Work Projects Administration; Be.05 U.S. Work Projects Administration U.S. Works Project Administration, St.01 Stanford University, Geology

UC (University of California), Be.01; Be.02; Be.03; Be.05; Be.06; Da.01; SF.02; SC.01; SC.03

———, Lick Observatory, Be.02; Be.05 Campbell | Drury | Galloway | Hill | Leuschner | Martens | Shane; SC.01; St.02 Brasch

UCB (University of California, Berkeley), Be.01; Be.02; Be.03; Be.05; Be.06; SF.02 J. Olmsted

———, Crocker Laboratory, Be.02 UCB; Be.05 Durbin | Hamilton | Lawrence | K. Scott

———, Donner Laboratory, Be.02 UCB; Be.05 Anger | Gofman | Grendon | H. Jones | Lawrence | Siri | Van Dyke

———, Lawrence Berkeley Laboratory (formerly the Radiation Laboratory), Be.01; Be.02 Childs | Lawrence | Research Corporation | UCB; Be.05 Alvarez | Birge | Brobeck | Calvin | Gofman | Goldhaber | Hamilton | Henderson | Huffman | Kamen | Lawrence | Livingood | Lofgren | Los Alamos | McMillan | Moyer | Seaborg | Thornton | Underhill | Woodyard | York

———, Lawrence Livermore Laboratory, Be.01; Be.02 UCB; Be.05 Crandall | Gofman | Los Alamos | Reynolds | York; St.03 Teller

———, Museum of Vertebrate Zoology, Be.05 Alexander | Grinnell Naturalists | Storer | Taylor

———, Radiation Laboratory. *See* UCB, Lawrence Berkeley Laboratory

UCD (University of California, Davis), Be.02; Be.05 Mrak | Spieth | Storer; Da.01

UCLA (University of California, Los Angeles), Be.05 Arnold | Beals | Knudsen | Libby | Mathias

UCSC (University of California, Santa Cruz), Be.02; SC.01; SC.03

UCSF (University of California, San Francisco), Be.02; Be.05 Meyer | Porter | K. Scott | R. Stone; SF.02

UNESCO, St.03 Intergovernmental Oceanographic Commission

Ulloa, Don Antonio de, Be.05 Ulloa

Underhill, Robert, Be.02 Underhill; Be.05 Underhill

Union Diesel Engine Company, Be.05 O. Fischer

United Aircraft Corporation, St.01 Siefert

University of Aberdeen, St.01 Findlay

University of Birmingham, Be.05 Boot; St.01 Kompfner

University of Chicago, Be.05 Arnold

University of Edinburgh, St.06 Black

University of Giessen, Be.05 Jaffé

University of Göttingen, Be.05 Bohr

University of Michigan, Be.05 Dana

University of Michigan, St.01 Webster

University of Minnesota, Be.05 Villars

University of Montpellier, St.02 Notebook

University of Oregon, Be.05 Moyer

University of Pennsylvania, St.06 B. Brown

University of Texas, Austin, Be.05 Herschel

University of the Pacific, Be.05 George; St.06 University of the Pacific

University of Tokyo, St.03 Tokyo Daigaku Statement

University of Washington, Be.05 Woodyard

Untermyer, Samuel, Be.05 Untermyer

Urey, Harold, Be.05 Arnold | Schulz

Urology, St.06 Cooper Medical College

Usinger, Robert L., Be.05 Usinger

Utility companies, Be.06. *See also* Power; Water resources

Vacuum tubes, Be.05 Clark | Heintz | Langmuir | Litton | Moore | Whinnery

Van Dyke, Donald C., Be.05 Anger | Van Dyke

Van Patten, Nathan, St.06 Van Patten

Vanderbilt Foundation, St.01 Vanderbilt Foundation

Varian Associates, Be.05 McCullough | Varian Associates; St.01 Hansen

Vasseur, Albert, Be.05 Vasseur

Vaucanson, J. de, St.02 Norman

Vaux, Henry J., Be.05 Vaux

Venus, transit of, Be.05 Pauly

Vesalius, Andreas, St.06 Spielmann

Villars, Donald S., Be.05 Villars

Virology, Be.05 Cutter Laboratories | Lennette | Stanley | R. Williams; St.06 Polio

Vitamins, Be.05 H. M. Evans

Viticulture, Be.02 UCB, Agriculture; Be.05 Amerine | Olmo; Da.01. *See also* Œnology

Vivisection, SF.02 California Society | Hooper Foundation. *See also* Antivivisection

Voorhies, Edwin, Da.01 Voorhies

Voy, C. D., Be.05 Voy

Walker, Harry, Da.01 Walker

War and science, Be.02 UCB, Donner Laboratory | UCB, Engineering, Science, and Management | Radiation Laboratory; Be.05 Boot | Durbin | G. Evans | Guerlac | Knudsen | Lawrence | Loeb | Los Alamos | Massachusetts Institute of Technology | National Research Council | Neyman | White; SF.02; St.01 Hansen | F. Terman St.03 Miller. *See also* Manhattan Engineer District; U.S. Army, Air Force, and Defense Department

Warren, Ambrose, Be.05 Newton

Warren, Earl, Be.05 Warren

Warren, John, St.06 Warren

Water resources, Be.05 Adams | Banks | Blaney | Bookman | R. Doble | Edmonston | Hinds | Langelier | Manson | R. Marshall | Mead | Rawn | F. Tibbetts; Be.06; Da.01 Gilmore; Sa.01; Sa.02 Givan | Hall | Natomas Company; St.01 Morris

Watkins, Lee H., Da.01 Watkins

Weakley, Charles Enright, St.03 Weakley

Weapons testing, St.01 Whitaker; St.03 Pictorial miscellany | Tokyo Daigaku Statement. *See also* War and science; Manhattan Engineer District; Radiation

Webster, David, Be.05 American Institute of Physics; St.01 Webster

Weeks, Walter, Be.05 Weeks

Weitbrecht, Robert, Be.05 Weitbrecht

Welch, Charles, Be.05 Welch

Welch, Marguerite, Be.05 Eastwood

Wesson, Miles, St.06 Wesson

West Bay Health Systems Agency, SF.02 West Bay

Western Union, St.02 Hyde

Weymouth, Frank, St.01 Weymouth

Wheeler, Helen-Mar, Be.05 Goodspeed

Whinnery, John, Be.05 Whinnery

Whitaker, Douglas, St.01 Whitaker

White, Harvey E., Be.05 White

Whitney, James P., St.06 Whitney

Whitney, Josiah D., Be.05 Brewer | Hoffmann | Voy

Wickson, Edward, Be.05 Wickson; Da.01 Wickson

Wiegand, Clyde, Be.05 American Institute of Physics

Wien, Wilhelm, Be.03

Wiggins, Ira, St.01 Wiggins

Wilbur, Dwight, St.01 D. Wilbur

Wilbur, Ray Lyman, St.01 R. Wilbur; St.06 Cooper Medical College | Reed | R. Wilbur

Williams, Arthur, Be.05 A. Williams

Williams, Robley C., Be.05 R. Williams

Wilson, Robert R., Be.05 American Institute of Physics

Winchester, Jonas, Sa.01 Winchester

Wines and vines, Da.01 Bayard | California wineries | Čebiš | Federal wine bottle labels | Wine newsletters | *Wines and vines*. *See also* Viticulture; Œnology

Winkler, Albert J., Be.05

Wise, W. Howard, Be.05 Wise

Witkin, Zara, St.03 Witkin

Woenne, Roy E., Be.05 Moore

Women and medicine, Be.05 Schlesinger Library | Medical College

Wong, B. C., Be.05 A. Williams

Wood, William, Be.05 Wood

Woodworth, Charles, Be.05 Woodworth

Woodyard, John, Be.05 Woodyard

Wooldridge, Dean E., Be.05 American Institute of Physics
Wright, William, SC.01 Wright
Wu, William, St.06 Wu
Wyckoff, Florence, Be.05 Wyckoff
Wyoming, Be.05 Aughey
Yale University, Be.05 Brewer
Yale, Charles, Be.05 Jepson
Yates, Lorenzo , Be.05 Camp | Yates
Yerkes Observatory, Be.05 Struve
Yerushalmy, Jacob, Be.05 Yerushalmy
Yoakum, Franklin L., Be.05 Yoakum
York, Herbert F., Be.05 York
Yoruba society, Be.05 Bascom
Young, Millard C., St.03 Young
Yurchenko, Ivan, St.03 Yurchenko
Yusufi, Kitab-i-Tib, St.06 Yusufi
Zeeman, Pieter, Be.03
Zimbardo, Phillip, St.01 Zimbardo
Zoology, Be.02 UCB, Vertebrate Zoology; Be.05 Alexander | S. Berry | Botta | Bryant | Chickering | R. Goldschmidt | Grinnell | Grinnell Naturalists | M. Hildebrand | S. Holmes | Kingsley | Kofoid | Leopold | Lombard | L. Miller | Needham | Ritter | Spieth | Storer | Taylor | Treadwell; Da.01 Storer; St.01 Berry | Gilbert. *See also* Ornithology

559056

3 1378 00559 0560

FOR REFERENCE

NOT TO BE TAKEN FROM THE ROOM

 CAT. NO. 23 012